T0298008

Clusterbean: Physiology, Genetics and Cultivation

Rakesh Pathak

Clusterbean: Physiology, Genetics and Cultivation

 Springer

Rakesh Pathak
Central Arid Zone Research Institute
Jodhpur, Rajasthan
India

ISBN 978-981-287-905-9   ISBN 978-981-287-907-3   (eBook)
DOI 10.1007/978-981-287-907-3

Library of Congress Control Number: 2015952529

Springer Singapore Heidelberg New York Dordrecht London

Printed on acid-free paper

Springer Science+Business Media Singapore Pte Ltd. is part of Springer Science+Business Media
(www.springer.com)

Dedicated to my parents Mrs. Urmila and Mr. R.S. Pathak, my wife Mrs. Sheela and my son Mayank

Preface

Clusterbean has become the most important export commodity in the farm sector. It is a drought tolerant leguminous crop of arid and semi-arid regions having a vast range of diverse and unique applications. It has a special place in the commercial scene due to the gum content in its seeds. India's share in the production of clusterbean seed is about 82 % and the country earns thousands of crores of rupees by exporting clusterbean products. Although the crop has attained the status of a number one agriculture export commodity in the farm sector, the productivity, production and expansion of the crop to non-traditional regions which are suitable for its cultivation require more attention. The clusterbean processing industry in India is fragmented; food safety alarms are growing in the export front because the technology of processing is not well developed. The high fluctuation in productivity and export are major concerns.

The present work provides in-depth information about the crop, its cultivation, genetic improvement, plant protection measures, management of abiotic stresses, molecular aspects, etc., and unites the value chain for the benefit of all the stakeholders of the crop, i.e. students, teachers, researchers and industrialists. This book is organized into seven chapters: Chapter 1 introduces the crop and gives the basic introduction, including prospects and constraints. Chapter 2 covers genetic improvement and variability of the crop and includes various tools and techniques used for the genetic improvement and creation of variability in the crop. Chapter 3 is dedicated to clusterbean gum and its by-products and covers the properties of gum and its application. Chapter 4 is devoted to the cultivation of the crop. Chapter 5 presents the plant protection aspect of the clusterbean. Chapter 6 addresses physiological and abiotic stress-related aspects of the crop. Chapter 7 covers the reviews of the genetic markers and the biotechnological works accomplished. The study has systematically referred to the research papers and data from industry and markets. It may prove a very useful book for industry as well as for the research community.

Rakesh Pathak

Contents

About the Author

Dr. Rakesh Pathak is a Senior Technical Officer at Central Arid Zone Research Institute, Jodhpur, Rajasthan and is a Fellow of Institution of Chemists (India). He obtained his doctoral degree on the phytochemical analysis of different cultivars of guar from the Jai Narain Vyas University, Jodhpur. Dr. Pathak has published more than 50 research papers and book chapters on molecular characterization and genetic diversity studies of several crops and trees in journals and books of national and international repute. He has molecularly characterized and submitted more than 150 novel gene sequences belonging to plant varieties/germplasm, macro-fungi, fungal pathogens and PGPRs with NCBI, USA, to address Indian biodiversity. Dr. Pathak has been regularly reviewing manuscripts as a referee for high-impact journals.

Chapter 1
Introduction

Abstract Clusterbean, tall and bushy annual herb having a deep rooted system, is a hardy and drought-resilient leguminous crop grown on sandy soils of arid and semi-arid regions. The Indian arid zone characterized by deficient moisture and nutrient, and high sunlight provides optimum agro-climatic conditions for the successful cultivation of clusterbean. It has been established as a high-valued cash crop in the arid and semi-arid regions due to its drought hardiness and multitude of usage and has occupied a special place in the commercial scene because of its gum.

1.1 Introduction

The clusterbean [*Cyamopsis tetragonoloba* (L.) Taub. (Syn. *C. psoraliodes*)], tall and bushy annual herb having a deep rooted system, is a hardy and drought-resilient leguminous crop grown on sandy soils of arid and semi-arid regions. It has been established as a high-valued cash crop in the arid and semi-arid regions due to its drought hardiness and multitude of usage and has occupied a special place in the commercial scene because of its gum. It is cultivated mainly in the rainy season and major producing states in India are Rajasthan, Haryana, Gujarat, Punjab and to a limited extent in Uttar Pradesh and Madhya Pradesh. The crop has now been a choice in southern India also. In addition to India, the crop is also grown in other parts of the world, viz. Sudan, Australia (Anonymous 1911), Brazil (Costa 1950), South Africa (Doidge 1952), Pakistan and parts of United States of America (Undersander et al. 1991).

The Indian arid zone characterized by deficient moisture and nutrient, and high sunlight provides optimum agro-climatic conditions for the successful cultivation of clusterbean, as the crop is known for high adaption towards poor and erratic rain, for its need of little surface water, abundant sunshine and low relative humidity during the cropping season (Pathak and Roy 2015). Gum obtained from clusterbean seeds is a choice of agrochemical in paper, food, mining, cosmetics,

© Springer Science+Business Media Singapore 2015
R. Pathak, *Clusterbean: Physiology, Genetics and Cultivation*,
DOI 10.1007/978-981-287-907-3_1

textile, oil and pharmaceutical industries across the world (Hymowitz and Matlock 1963; Pawlik and Laskowski 2006; NRAA 2014). There are number of cluster-bean varieties grown in India for different purposes. The major varieties are for vegetable, forage, fodder, cover crop and seed gum types. Indian cultivars and germplasms have wide variability for morphological and agronomic variability, i.e. pubescence of the plant, pattern of branching, bearing habit, shape, size and texture of the pods, seed size, colour and quantity of gum in seeds (Dabas et al. 1995).

The world's total clusterbean production has been figured around 7.5–10 Lakh tonnes every year. The production list of clusterbean is dominated by India as leading producers of the crop in the world contributing to around 75–82 % of the total production. Whereas, Pakistan follows India in the list with 10–15 % share in the world's total production. The consumption pattern of its seed is largely influenced by the demands from the petroleum industries in USA and oil fields in the Middle East. The trend of consumption has also increased in rest of the world that has led to its introduction in many countries. The world market of the crop is estimated more than 1.5 lakh tonnes annually. The main importer countries of clusterbean gum are Australia, Austria, Brazil, Canada, China, Chile, Germany, Greece, Ireland, Italy, Japan, Mexico, Portugal, Sweden, UK and USA (NRAA 2014).

Rajasthan occupies the largest area (82.1 %) under cultivation of clusterbean followed by Haryana, Gujarat and Punjab. Clusterbean was also grown regularly in Uttar Pradesh, Madhya Pradesh and Odisha during 1970s but due to closing of the processing facilities in Uttar Pradesh and Madhya Pradesh, the cultivation in these states is now insignificant. Rajasthan is the largest clusterbean producing states in the world as it dominates the Indian production scenario contributing to 70 % of the total production in India followed by Haryana (12 %) and Gujarat (11 %). In Rajasthan, Churu, Bikaner, Jaisalmer, Barmer, Nagaur, Hanumangarh, Jodhpur, Shriganganagar, Jaipur, Sirohi, Dausa, Jhunjhunu and Sikar are the major clusterbean producing districts whereas Bhiwani, Gurgaon, Mahendragarh and Rewari are the main districts of Haryana involved in the clusterbean production. In Gujarat Kuchchh, Banaskantha, Mehasana, Sabarkantha, Vadodara and Ahmedabad are the major clusterbean producing districts. After seeing great revenues with the crop during previous years by Rajasthan farmers, farmers in Ananthapur, Guntur, Karnool, Karimnagar, Nellor, Prakasam and Rangareddi districts of Andhra Pradesh have also started the cultivation of this crop for seeds in more than 1000 ha (NRAA 2014). Clusterbean has occupied its place in the Indian commodity exchanges like National Commodity Derivatives Exchange Ltd., Multi Commodity Exchange of India Ltd., etc.

The dicotyledonous seed of clusterbean from outer side to the interior consists of three major fractions, viz. the husk or hull (14–17 %), endosperm (35–42 %) and germ or embryo (43–47 %). Green, matured pod, seed and a cross section of the seed and its constituents are given in Fig. 1.1 and Table 1.1, respectively.

The clusterbean seed has rather a large endosperm unlike most of the other legumes. The dull, white-coloured and unwrinkled seeds of clusterbean are preferred for gum processing and black seeds are believed to be of low quality (Bhatia et al.

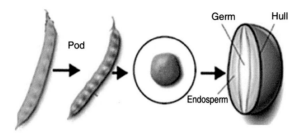

Fig. 1.1 Cross section of clusterbean seed

Table 1.1 Constituent of clusterbean seed

Part of seed	Protein (%)	Ether extract (%)	Ash (%)	Moisture (%)	Fibres (%)
Hull	5.0	0.3	4.0	10.0	36.0
Endosperm	5.0	0.6	0.6	10.0	1.5
Germ	55.3	5.2	4.6	10.0	18.0

Source Vishwakarma et al. (2009); http://www.dsir.gov.in/reports/tmreps/guar.pdf

1979; Hymowitz and Matlock 1967). But studies suggest that the black-coloured seeds may be recommended for planting or gum extraction with little loss in stand or gum yield (Bhatia et al. 1979). The galactomannan is found in the endosperm which makes up about 35 % of the dry weight of the seed, 80–90 % being pure galactomannan, having 1:2 ratio of galactose to mannose (Das and Arora 1978).

1.2 Origin

Clusterbean has been grown in India since ancient time for vegetable, manure and fodder purposes. The presence of a number of wild relatives of clusterbean in Africa suggests that it was most probably originated in Africa (Gillette 1958). It is possible that clusterbean was domesticated very early in the Africa and Arabia and made its way to Indo-Pakistan subcontinent. On the other hand, Whistler and Hymowitz (1979) mentioned that the name of cultigen in Arabic *hindia* suggests it to be an Indian origin. However, a trans-domestication concept proposed by Hymowitz (1972), illustrates that a drought-tolerant *C. senegalensis* reached to Indo-Pakistan subcontinent via Senegal to Saudi Arabia from the semi-arid and savannah zone of Sahara as waste material during Arab-Indian trade.

Hymowitz (1972) hypothesized that *C. senegalensis* the wild progenitor of clusterbean may have passes the Ethiopian route and then been carried as horse fodder to India where it was domesticated. As it is well-known that horses were the major trade between the Arabs and Indians, there is a possibility that Arabs have boarded their ships in large quantities of fodder of clusterbean to feed their horses. The plants of *C. senegalensis* probably were cut and carried along the ship as fodder. Since the climatic conditions in the Indo-Pakistan subcontinent were favourable to *C. senegalensis*, seeds get germinated and became the basis of clusterbean. According to an another hypothesis, when the trade between middle east and India flourished during the silk route trading period, it is believed that traders brought clusterbean pods with them and wherever they went the seeds of the crop got scattered resulting in the introduction of the crop. These hypotheses on the origin of the clusterbean appear quite speculative. Except taxonomic studies, no detailed molecular or genetical studies are available in the literatures to validate the claim or prove the hypotheses.

Chavalier (1939) postulated that *C. senegalensis* probably extended up to Sindh where after domestication a few of its cultigens became cultivated in India whereas Vavilov (1951) suggested that India is the geographical centre of clusterbean variability. Dabas and Thomas (1986) indicated that the clusterbean perhaps has been domesticated in the western Rajasthan. Hymowitz (1972) believed that the African wild species *C. senegalensis* appeared to be the ancestor of West African *C. tetragonoloba*.

1.3 Taxonomic Classification

Domain: *Eukaryota*
Kingdom: *Plantae*
Subkingdom: *Viridaeplantae*
Phylum: *Magnoliophyta*
Subphylum: *Euphyllophytina*
Infraphylum: *Radiatopses*
Class: *Magnoliopsida*
Subclass: *Rosidae*
Superorder: *Fabanae*
Order: *Fabales*
Family: *Fabaceae*
Subfamily: *Paplionaceae*
Tribe: *Indigofereae*
Genus: *Cyamopsis*
Specific epithet: *tetragonoloba*—(L.)Taub.

1.4 Botany

Clusterbean belongs to the family Fabaceae or Leguminaceae and subfamily paplionaceae. It is a self-pollinated crop with $2n = 14$ chromosomes (Hymowitz and Upadhya 1963). Gillette (1958) divided the genus *Cyamopsis* in to three races, viz. *C. tetragonoloba* (L.) Taub, *C. senegalensis* Guill. and Perr. and *C. serrata* Schinz. The haploid and diploid chromosome number of all the three genus species of *Cyamopsis* were reported to be n = 7 and 2n = 14. The status of intermediate forms between *C. serrata* and *C. senegalensis* was left unsettled; although he suggested that the intermediate form may be the result of hybridization between the species. Later on Torre (1960) recognized this intermediate form as a distinct species and named as *C. dentate* (N.E. Br.) Torre. Sen (1938) noted a distinct similarity between the genus *Cyamopsis* and *Indigofera* and hypothesized that the genus *Cyamopsis* was a derived aneuploidy from *Indigofera* but Gillette (1958) preferred to retain *Cyamopsis* as a distinct genus.

Menon (1973) studied the morphology of species *C. senegalensis* and *C. serrata* and found *C. senegalensis* as a slow-growing annual herb with narrow pentafoliolate leaves and small pods having white seeds. The species *C. serrata* is also a slow-growing annual having narrowed trifoliolate leaves and pink-coloured seeds. *C. dentata* is an erect branching annual herb, 30–35 cm tall, over-clothed with oppressed biramous hairs with entire or dentate leaflets having mauve purple-coloured flowers, glabrous and erect pods and is found in the same habitats as *C. senegalensis* and *C. serrata* (www.villw.ge.ch/musinfo/bd/cjb/africa/details.php?langue=an&id=127187).

Clusterbean is generally 50–150 cm tall annual herb having long tap root and well-developed laterals with rhizobium nodules. Branching habit in clusterbean may be erect, basal branching and branched. The erect category of the crop has predominantly zero to two branches, in case of basal branching three or more branches are present at the base of the plant while in branched category branching occurs in the plant along with the main stem. It bears 4–10 branches in case of branched cultivars while unbranched cultivars have main stem only, which is heavily clustered with pods. The leaves are medium-sized, alternating trifoliate, pubescent or glabrous, borne on the long petiole and the stem is tall and slender. Clusterbean produces 8–9 mm long purplish-pink or white-coloured flowers. *C. senegalensis* and *C. serrata* have shorter plant height (about 30 cm) with smaller leaves and pods when compared with *C. tetragonoloba*. There are several pods/cluster, with a mean of nine seeds per pod. The seed of *C. senegalensis* and *C. serrata* are smaller and 100-seed weight varies between 1 and 1.4 g. The seeds of these species are short and cylinder-shaped, while seeds of *C. tetragonoloba* are nearly round. Pod shattering is also observed in *C. senegalensis* and *C. serrata* at maturity which is not noticed in *C. tetragonoloba* (Menon 1973).

1.5 Floral Characteristics

Clusterbean is completely self-fertile and self-pollinated crop because of its cleistogamous nature. The extent of out crossing has been found to vary from 0.3 to 7.9 % (Chaudhary and Singh 1986; Ahlawat et al. 2012). The flowers borne in the axillary racemes are bisexual, almost sessile and about 9 mm long. There are about 50 flowers in each inflorescence and approximately 12.5 % of them develop in to pods. It is reported that an initiated bud requires 35 days to develop into flower (Menon 1973). The inflorescence is a raceme of about 9–13 cm long in the branched type and 15–20 cm long in the erect or sparsely branched types. Normally 40–60 flowers are present in branched types and 50–70 in erect sparsely branched inflorescence (Menon 1973). The calyx has unequal linear teeth-type 5 sepals and the corolla has 5 petals. The standard is circular, the wing petals are oblong and keel petals are as long and broad as the wing petals. The ovary is linear, sessile and one-celled containing 6–10 ovules. The style is short and slender while stigma is head-shaped (Menon et al. 1968; Ahlawat et al. 2012). Flowers pass through an array of colours from white to deep blue from the bud stage to petal drop. A mature bud is creamy white in colour, which changes to light pink or white. Petals develop a pink colour just prior to opening. The flower colour in general is purplish to pink. However, white colour flowers are also seen on some cultivars. There are 10 stamens in diadelphous condition having opercular anthers. Anthers dehisce within 1½ to 2 h before flower opening and pollen was found to be capable of germination 2 h, before and 11 h after the opening of the flower. The flower opens early in the morning and shed petals the same day, while late opening flowers shed petals in the next morning (Menon 1973).

The longevity of pollen at room temperature varied from 11–13 h with maximum pollen germination of 46 %. Each pollen cell normally produced one tube, but two tubes are also observed with a length of 1.5 mm in some cultivars Menon et al. (1968). Ahlawat et al. (2012) studied the biochemical composition of stigma and style in *Cyamopsis* spp. and reported that the protein content of stigma and style was nearly identical in all the three species, while total soluble carbohydrate content in *C. tetragonoloba* and *C. serrata* was nearly identical, while it was low in *C. senegalensis*.

1.6 Cytogenetics

Ayyangar and Krishnaswami (1933) observed seven pairs of chromosomes in clusterbean, later it was confirmed by number of other investigators (Ahlawat et al. 2012; Mullainathan et al. 2014). Sandhu (1988) demonstrated that *C. serrata* also had seven pairs of chromosomes. It has been noticed that in spite of a general similarity in the karyotype, the species *C. serrata* has chromosome complement longer than the rest of species. Among the species, it would be reasonable to assume in view of slightly longer chromosomes that *C. serrata* may represent a comparatively primitive status in relation to the rest of the species. Since no secondary constrictions or subterminal or terminal centromeres were observed in any of the species, it appears that unequal translocations and other processes might have played a role in change of karyotypes in *Cyamopsis*. The karyomorphological variation in chromosome length, sub-median primary constriction and a pair of chromosomes has been reported in clusterbean (Purohit et al. 2011). The karyotypic studies were conducted by Arora et al. (1985) among *C. senegalensis*, *C. serrata* and cultivated species *C. tetragonoloba* and showed an identity in chromosome number (2n = 14) and similarity in morphology. The chromosomes were found to be medianly or sub-medianly constricted. The chromosome complement of *C. serrata* was longer than *C. senegalensis* and *C. tetragonoloba* and indicates that it may represent a comparatively primitive status in relation to rest of the *Cyamopsis* species. The chromosome length in *C. tetragonoloba, C. serrata* and *C. senegalensis* ranged between 7.16 and 9.26 μm, 7.33–11.52 μm and 5.50–9.60 μm, respectively. The total form percent value of *C. tetragonoloba* (44 %) showed the close proximity with that of *C. serrate* (45 %) and *C. senegalensis* (45 %) indicating the similarity in karyotypes among these species (Arora et al. 1985).

C. *senegalensis* is regarded as the ancestral form of clusterbean and has similar gum concentration, composition and viscosity characteristics (Strickland and Ford 1984) and has close affinity with *C. serrata* than with *C. tetragonoloba* based on allozymes, RFLP and RAPD profiles (Hirematha et al. 1996). Weixin et al. (2009) also revealed that *C. senegalensis* is more closely related to *C. serrata* than to *C. tetragonoloba* indicating that *C. tetragonoloba* is distinct from both *C. senegalensis* and *C. serrata*.

1.7 Molecular Biology

Clusterbean is comparatively less exploited for establishment of molecular marker to assess its phenotypic database however, the isozyme diversity in relation to domestication (Mauria 2000) and genetic variability (Brahmi et al. 2004) has been carried out. Few studies on different PCR-based molecular markers techniques, viz. RAPD, ISSR, EST, SCAR, CAPs, ITS, SSRs have been used for various purposes in clusterbean. Further, the transmission of transgene in clusterbean has also been studied and aberrant transmission at a high frequency in the process of going from the primary transformants to the first offspring generation was recorded within the major parts of the transgenes (Joersbo et al. 1999). Clusterbean poorly responds to the in vitro culture of endosperm, genetic transformation and subsequent regeneration of the transformed tissue (Prem et al. 2005; Verma et al. 2013). Molecular documentation of clusterbean germplasm is an urgent need for conservation and assessment of variability intended for morphophysiological and industrial qualities.

1.8 Uses

Clusterbean is multipurpose crop. Its green pods are used as vegetable, grains as pulses and green plants as fodder and for soil manuring purpose. The entire plant is useful and nothing goes waste. On an average basis, it has been estimated that clusterbean may fix nearly 30 kg nitrogen/ha. The leaves of the crop completely shed up at maturity and add organic carbon to the soil. In this way the crop is a good source to take care of soil health. The crop produces seed that contain a galactomannan gum for which a multitude of food and non-food application have been developed and an enormous range of value added products/derivatives have appeared in the market with more than one lakh patents globally, establishing it as an important commercial crop (APEDA 1999).

Clusterbean gum in solution behaves as an excellent thickener and its thickening power is 10 times higher than that obtained from starch. It is stable over a wide range of pH and improves flowability and pumpability of the fluid. The use of clusterbean gum can be envisaged in the system where water is an important factor. The by-product of guar gum industry consisting of the outer seed coat and germ portion contained about 35–47.5 % crude protein, termed as guar meal, is a potential source of protein, nutrients and fibres with high digestibility. It is used as a feed for livestock including poultry and fishery Arora (1981).

The gum of clusterbean may promote a protective, pain-relieving and healing effect on gastric ulcers, contribute to lowering cholesterol, blood pressure and blood sugar levels and also play a positive role in general weight loss and obesity with the promotion of fullness and appetite suppression (Sharma et al. 2011; James 2002). The methanolic extract of clusterbean seeds has anticancer activities (Badr et al. 2014), seed are laxative and boiled seeds are used to cure plague, inflammation, sprains (Khare 2004), arthritis (Katewa et al. 2004), as anti-oxidant, anti-bilious and liver enlargement. It is also a popular choice in the manufacture of skin creams and gels to thicken and stabilize preparations. Certainly, it is a natural and affordable plant based alternative to be considered for various uses.

1.9 Prospects

Clusterbean is a unique plant and every part of the plant can be used for different applications. It provides a bright prospect to the growers, consumers, environmentalists and industrialists. With its advantages the crop has become an important commodity in the commodity exchange and it has become the highest foreign exchange earner of Rs. 212.87 billion among all the agricultural export commodities followed by Basmati rice in the recent year of 2012–13 (NRAA 2014). The market arrival and prices of seed and gum vary from year to year depending on the monsoon conditions. Presently about 12 leading processors, manufacturers or companies are involved and some of them are getting into contact or contract farming with a large number of small holders. An outline of the prospect of the crop is underlined as follows:

1. The crop is a choice of cash crop in the arid and semi-arid region which gives adequate production without any serious care even in harsh environmental conditions. Clusterbean utilizes heat and moderate salinity and good production

can be realized. It has an ability to combat desertification and improve soil environment.

2. The long tap root system of the crop extract moisture from deeper layers of soil leading to the cooling of canopy.
3. Meristematic type of rhizobium nodules associated with roots of the cluster-bean provides greater recovery potential to the crop and helps in fixing the atmospheric nitrogen in adequate quantity.
4. The young and tender pods used for human consumption are good source of vitamin A, carbohydrate, vitamin C, iron, calcium, phosphorus and energy. Being a good source of nutrition, it is a choice of vegetable for human consumption.
5. A large area of arid and semi-arid regions are devoted to clusterbean farming during rainy season (*kharif*) and number of gum industries have grown in this region. In this way, a sizable number of jobs are offered to the people of this region due to this crop.
6. The gum and meal obtained from the clusterbean seed have established a great catch and source of foreign exchange.
7. The crop has attained a respectable place among the resource-constraint farmers depending on marginal lands for their livelihood.

1.10 Constraints

It has been established that clusterbean is a versatile crop and being used in various ways. But it is facing number of constraints which affect its adequate production to cope the requirement. It may be summed up as follows:

1. The productivity of the crop fluctuates over the years. Varieties having stable yield is unavailable.
2. Absence of varieties susceptible for Alternaria leaf spot, bacterial blight, root rot, wilts and anthracnose diseases.
3. Absence of early maturing varieties (<80 days) having drought-escape mechanisms with high seed yield and gum content.
4. Clusterbean is always given second priority by the farmers of arid region. Mostly it is planted in the poor fertility and poor textured soil. Marginal status plays major constraint for the crop.
5. Support price for this crop is unavailable. A suitable market for sale of produce is limited.
6. Most of the available varieties are not coupled with higher yield and gum content.
7. Cheaper alternatives for clusterbean gum are being looked out due to price hike and unavailability of seeds.

References

Ahlawat A, Dhingra HR, Pahuja SK (2012) Biochemical composition of stigma and style in *Cyamopsis* spp. Forage Res 38(1):53–55

Anonymous (1911) Guar. Agricultural Gazette, New South wails 22:1000

APEDA (1999) A study on guar gum, p 121

Arora SK (1981) Guar, its uses are many and expanding all the times. In: 4th ICAR guar research and development workshop held at CAZRI, Jodhpur on March 10–11

Arora RN, Sareen PK, Saini ML et al (1985) Karyotype analysis in three species of genus *Cyamopsis*. Indian J Genet 45(2):302–309

Ayyangar GNP, Krishnaswami N (1933) A note on the chromosome number in clusterbean (*Cyamopsis psoralioides* DC). Indian J Agric Sci 3:934–935

Badr SE, Abdelfattah MS, El-Sayed SH et al (2014) Evaluation of anticancer, antimycoplasmal activities and chemical composition of Guar (*Cyamopsis tetragonoloba*) seeds extract. Res J Pharm Biol Chem Sci 5(3):413–423

Bhatia IS, Nagpal ML, Singh P et al (1979) Chemical nature of the pigment of the seedcoat of guar (Cluster Bean *Cyamopsis tetragonolobus* L. Taub). J Agric Food Chem 27:1274–1276

Brahmi P, Bhat KV, Bhatnagar AK (2004) Study of allozyme diversity in guar (*Cyamopsis tetragonoloba* L. Taub.) germplasms. Genet Resour Crop Evol 51:735–746

Chaudhary BS, Singh VP (1986) Extent of outcrossing in guar (*Cyamopsis tetragonoloba* (L.) Taub.). Genet Agrar 34:59–62

Chavalier A (1939) Research on species of genus *Cyamopsis* forage plants for tropical and semi arid countries. Rev Bot Appl 19:242–249

Costa AS (1950) Beta pettalaris, a test plant for tobacco white necrosis virus. Bragantia 10:275–276

da Torre AR (1960) Texa angolensia nova vel minus cognita-1. Men Junta Invest Ultram Ser 19:23–66

Dabas BS, Thomas TA (1986) Shattering guar and its significance. Int J Trop Agric 2:185–187

Dabas BS, Phogat BS, Rana RS (1995) Genetic resources of clusterbean in India. In: Sharma B (ed) Genetic research and education: current trends and the next 50 Years. Indian Society of Genetics and Plant Breeding, New Delhi, pp 63–70

Das B, Arora SK (1978) Guar seed its chemistry and industrial utilization of gum. In: Guar its improvement and management. Forage Res 4:79–101

Doidge EM (1952) South African funfi. *Pretoria National Herbm*

Gillette JB (1958) *Indigofera* (Microcharis) in tropical Africa with the related genera *Cyamopsis* and *Rhyncotropis*. Kew Bull Add Ser 1:1–66

Hirematha SC, Ramamoorthy J, Cai Q et al (1996) Analysis of genetic relationships in the genus *Cyamopsis* (Fabaceae) using allozymes, RFLP and RAPD markers. Am J Bot 83:207

Hymowitz T (1972) The trans domestication concept as applied to guar. Econ Bot 26:49–60

Hymowitz T, Matlock RS (1963) Guar in the United States, Oklahoma Agri Expt Station. Tech Bull 611:1–34

Hymowitz T, Matlock RS (1967) Variations in seed coat colour of 'groehler' guar [*Cyamopsis tetragonolobus* (L.) Taub.]-genetic or environmental? Crop Sci 1:465

Hymowitz T, Upadhya MD (1963) The chromosome number of *Cyamopsis serrata* Schinz. Curr Sci 32:427–428

James AD (2002) Handbook of medicinal herbs. CRC Press, Washington DC, pp 118–119

Joersbo M, Brunstedt J, Marcussen J et al (1999) Transformation of the endospermous legume guar (*Cyamopsis tetragonoloba* L.) and analysis of transgene transmission. Mol Breed 5:521–529

Katewa SS, Chaudhary BL, Jain A (2004) Folk herbal medicine from tribal areas of Rajasthan. Indian J Ethanopharmacol 92:41–46

Khare CP (2004) Indian herbal remedies. Springer, New York 525p

Mauria S (2000) Isozyme diversity in relation to domestication of guar [Cyamopsis tetragonoloba (L.) Taub.]. Indian J Plant Genet Resour 13:1–10

Menon U (1973) A comparative review on crop improvement and utilization of clusterbean(*Cyamopsis tetragonoloba* (L.) Taub). Department of Agriculture, Rajasthan, Jaipur. Monogr Ser 2:1–51

Menon U, Radhare NS, Bhargava PD (1968) Pollen studies in guar (*Cyamopsis tetragonoloba* (L.) Taub.). Indian J Palynol 4:51–53

Mullainathan L, Aruldoss T, Velu S (2014) Cytological studies in cluster bean by the application of physical and chemical mutagens (*Cyamopsis tetragonoloba* L.). Int Lett Nat Sci 16:35–40

NRAA (2014) Potential of rainfed guar (clusterbeans) cultivation, processing and export in India. Policy paper No. 3 National Rainfed Area Authority, NASC Complex. DPS Marg, New Delhi-110012, India, 109 p

Pathak R, Roy MM (2015) Climatic responses, environmental indices and interrelationships between qualitative and quantitative traits in Clusterbean (*Cyamopsis tetragonoloba* (L) Taub.) under arid conditions. Proc Natl Acad Sci, India, Sect B Biol Sci 85(1):147–154

Pawlik M, Laskowski JS (2006) Stabilization of mineral suspensions by guar gum in potash ore flotation systems. Can J Chem Eng 84:532–538

Prem D, Singh S, Gupta PP et al (2005) Callus induction and de novo regeneration from callus in guar (*Cyamopsis tetragonoloba*). Plant Cell Tissue Organ Tissue 80:209–214

Purohit J, Kumar A, Hynniewta M et al (2011) Karyomorphological studies in guar (*Cyamopsis tetragonoloba* (Linn.) Taub.)-An important gum yielding plant of Rajasthan, India. Cytologia 76(2):163–169

Sandhu HS (1988) Interspecific hybridization studies in genus *Cyamopsis*. Ph.D. Thesis (Unpublished), CCS HAU, Hisar

Sen HA (1938) Chromosome number relationship in the leguminosae. Bibl Genet 12:175–336

Sharma P, Dubey G, Kaushik S (2011) Chemical and medico-biological profile of *Cyamopsis tetragonoloba* (L) Taub: an overview. J Appl Pharma Sci 1(2):32–37

Strickland RW, Ford CW (1984) *Cyamopsis senegalensis*: potential new crop source of guaran. J Aust Inst Agric Sci 47–49

Undersander DJ, Putnam DH, Kaminski AR et al (1991) Guar. In: Alternative field crops manual, University of Wisconsin-Madison. www.hort.purdue.edu/newcrop/afem/guar.html

Vavilov NI (1951) Phytogeographic basis of plant breeding: the origin, variation, immunity and breeding of cultivated plants. Chronica Bot 13:1–366

Verma S, Gill KS, Pruthi V et al (2013) A novel combination of plant growth regulators for in vitro regeneration of complete plantlets of guar *Cyamopsis tetragonoloba* L. Taub]. Indian J Exp Biol 51:1120–1124

Vishwakarma RK, Nanda SK, Shivhare US et al (2009) Status of post harvest technology of guar (*Cyamopsis tetragonoloba*) in India. Agric Mechanization Asia Afr Lat Am 40(3):65–72

Weixin L, Anfu H, Peffley EB et al (2009) Genetic relationship of guar commercial cultivars. Chin Agric Sci Bull 25(2):133–138

Whistler RL, Hymowitz T (eds) (1979) Guar: agronomy, production, industrial use and nutrition. Purdue University Press, W. Lafayette, Indiana, p 124

Chapter 2
Genetic Improvement and Variability

Abstract Clusterbean has been reported with vast variability including branched or unbranched plant types, hairy or smooth stems, straight or sickle-shaped pods, pubescent or glabrous leaves, determinate or indeterminate growth, regular or irregular pod bearing habits. Various methods used for genetic improvement of the crop and for assessment of variability are discussed in this chapter.

2.1 Introduction

The use of diverse germplasms in crop improvement is one of the most sustainable methods to conserve valuable genetic resources and simultaneously to increase agricultural production and food security (Ogwu et al. 2014). The genetic resources have been used as foundations to broaden adaptation, to create resistance against disease, insect, stress, to improve quality, stature and to increase yield potential of the crop. The wild relatives are the major source of genetic support and help the crops in maintaining its valuable status. The selections made through local land races have resulted in the release of superior genotypes of clusterbean (Henry et al. 1992; Bharodia et al. 1993; Mishra et al. 2009).

2.2 Genetic Variability

The genetic variability within the crop helps in its maximum utilization and becomes the basis of species conservation. Similarly, the knowledge of genetic divergence among the varieties has immense importance for plant breeders while use of diverse germplasm has a significant contribution towards high yield and quality characteristics (Maniee et al. 2009; Singh et al. 2014). Clusterbean germplasm has been reported with vast variability including branched or unbranched plant types, hairy or smooth stems, straight or sickle-shaped pods, pubescent or

© Springer Science+Business Media Singapore 2015
R. Pathak, *Clusterbean: Physiology, Genetics and Cultivation*,
DOI 10.1007/978-981-287-907-3_2

glabrous leaves, determinate or indeterminate growth, regular or irregular pod bearing habits (Saini et al. 1981). Variability for different characters in cluster-bean has been summed by Mishra et al. (2009), the ranges described are: days to 50 % flowering (25–76 days), days to 50 % maturity (66–128 days), plant height (55–238 cm), branches/plant (0–29), clusters/plant (2–86), pod length (1.6–17 cm), pods/plant (4–412), seeds/pod (2–15), seed yield/plant (1.2–71 g) and 100-seed weight (1.5–5.3 g). Pathak et al. (2011c) reported wide range of variability for different morphological traits of clusterbean, viz., plant height (31.8–42.8 cm), number of primary branches (7.7–13.1), number of secondary branches (10.8–29.8), number of pods/plant (21.1–44.9), number of seeds/pod (6.9–9.4), 100-seed weight (2.57–3.06 g), seed yield/plant (5.5–11 g) and days to 50 % flowering (30.3–35.8 days). Similarly, wide variation in different biochemi-cal parameters of clusterbean seed, viz., endosperm (30.4–46.3 %), gum con-tent (23.5–33.5 %), crude fibre (4.1–8 %), fat content (1.8–5.2 %), crude protein (28.3–35 %), ash content (3.5–6 %) and carbohydrate content (38.8–59.1 %) has also been reported (Pathak et al. 2011a).

The inheritance studies indicates that the foliage colour, branching and pol-len fertility are controlled by a single gene with the dominance of the dark green foliage colour over the light green colour, unbranching over branching and pol-len fertility over pollen sterility, whereas, pod length was controlled by several gene pairs (Dabas 1975). Singh et al. (1990a) reported dominance of serrated leaf margin over the smooth margins and pubescent leaf surface over the glabrous sur-faces. Saharan et al. (2004) observed the dominance of hairy types over non-hairy, branched over unbranched, purple colour flower over white colours, broad leaf over narrow leaves, curved pods over non-curved pods and plucked leaf surface over non-plucked leaf surface. The inheritance of branching pattern and leaf serra-tion in clusterbean was also studied and it was observed that single dominant gene was responsible for branching pattern and leaf serration (Saharan et al. 2004).

Understanding of the inheritance of gum content is a key for successful genetic improvement of clusterbean for quality traits. The inheritance of gum content in the seed of clusterbean is quite complex in nature and both additive and non-addi-tive types of gene action are operating in the expression of gum content (Dabas 1975). The gum content expression is reported to be controlled by additive, domi-nant and epistatic effects and modified by the environment. Additive gene effects are more important in most of the crosses, therefore, it is suggested that selective intermating in F_2 or F_3 generation should be resorted to exploit additive type of genetic components and plant or lines may be selected in later generations for genetic improvement for higher gum content (Singh et al. 1990b). The operation of both the additive and non-additive gene actions in the expression of gum con-tent, its positive association with seed yield and negative correlation with seed weight (Pathak et al. 2011a) have hampered the breeding of clusterbean for gum content, therefore, some concrete efforts are required to be taken up for improve-ment of gum and protein content.

Burton and De Vane (1953) suggested that Genotypic Coefficient of Variation (GCV) together with heritability estimates would give reliable indication about the expected improvement of desired trait. Clusterbean has been reported with moderate-to-high estimates of GCV for seed yield and yield attributing traits (Pathak et al. 2011a; Kapoor and Bajaj 2014). The seed yield/plant is a primary trait which depends on number of variables, but high GCV values reported for a particular trait indicate that direct selection for seed yield is one of the best approaches for its genetic improvement. Higher estimates of variability observed in the characters show that there is ample scope for selection to improve the traits in clusterbean (Arora and Gupta 1981).

In quantitative genetic analysis of clusterbean, Chaudhary et al. (2003) found higher estimates of GCV for number of clusters/plant, primary branches/plant, pods/plant, seed yield/plant, biological yield/plant and days to 50 % flowering. They found similar trends for Phenotypic Coefficient of Variation (PCV). Reflecting the susceptibility to environmental fluctuation they observed wide differences between PCV and GCV for number of clusters/plant, number of pods/plant, seed yield/plant, biological yield/plant, harvest index and seeds/pod.

The study of heritability in conjunction with genetic advance is more useful than heritability alone in prediction of resultant effect of selecting the best individual (Singh et al. 2010). Traits having high heritability and genetic advance are supposed to be under control of additive genes, hence the characters with higher heritability and genetic advance may be improved by selection based on phenotypic performance (Shekhawat and Singhania 2005). The high estimates of heritability and genetic advance for various traits of clusterbean indicate that the seed yield and its components are governed by non-additive gene action (Pathak et al. 2011c), whereas days to flower initiation, plant height, number of branches/plant, number of pods/plant, number of seeds/pod are governed by additive gene action (Arora et al. 1999). Studies revealed that both the additive and non-additive genes have important role in the expression of almost all the traits in clusterbean. Heritability was high for plant height and number of clusters/plant and moderate in 100-seed yield and number of clusters/plant. Comparatively, high value of heritability coupled with genetic advance was observed for number of branches, plant height and number of clusters/plant. Similarly, moderate-to-high genetic variability and heritability estimates for number of traits among the varieties of clusterbean has been reported (Anandhi and Oommen 2007; Weixin et al. 2009).

The characters, viz., number of branches/plant, leaf area index, number of pods/plant, seed yield/plant, number of clusters/plant, harvest index, number of pods/cluster and plant height showed higher estimates of GCV and PCV (Raghu Prakash et al. 2008). Stafford and Barker (1989) reported heritability and interrelationships of pod length and seed weight in clusterbean and suggested that the selection for gum content can expect progress, but not as quickly as other traits, such as pod length and seed weight. Rai et al. (2012) observed maximum range of variability for number of branches, plant height, clusters/plant, pod length and pod yield/plant. High heritability coupled with high genetic gain in percentage was observed for pod yield/plant, number of pods/plant, days to 50 % flowering, number of branches and plant height. High heritability coupled with low genetic advance was recorded for pods/cluster, number of seeds/pod and pod width. Girish et al. (2013) reported broad genetic base among the genotypes on the basis of higher value of GCV and PCV for stem girth, vegetable pod yield/plot, dry pod yield/plot, seed yield/plant, endosperm and gum content. They reported high heritability coupled with high genetic advance for stem girth, cluster length, pod length, endosperm, protein and gum content. Similarly, Kapoor and Bajaj (2014) also recorded high heritability along with high genetic advance for number of leaves/plant, number of branches, dry matter yield and green fodder yield. Vir and Singh (2015) estimated the genetic variability, intercharacters associations in the germplasm of clusterbean and observed high degree of genetic variability for seed yield, 100-seed weight, number of seeds/pod, number of pods/plant, number of pods/cluster, number of branches/plant, number of clusters/plant, plant height, days to 50 % flowering and days to maturity during both summer and rainy seasons.

2.3 Karyotype Analysis

Determination of chromosomes numbers and karyotype analysis is an important method for the assessment of the genomic status of morphologically diverse populations. Meagre information is available on these aspects in clusterbean as there are difficulties in getting well spread mitotic metaphase chromosome preparations, staining and resolution resulting problems for identification of correct position of centromere(s) (Purohit et al. 2011). The somatic chromosome counts reported by many authors for clusterbean is $2n = 14$ (Jahan et al. 1994; Patil 2004). Purohit et al. (2011) reported the $2n$ number of chromosome as 14 in the root-tip cells without any indication of existence of polyploidy/aneuploidy or any numerical variation.

Patil (2004) has guessed that the karyotypes of *C. senegalensis* and *C. serrata* could be resolved into asymmetrical and symmetrical types, respectively. *C. serrata* has slightly longer chromosomes and represents comparatively primitive status in relation to the rest of the species on the basis of karyotypic studies. Arora et al. (1985) suggested the presence of unequal translocations in chromosomes that plays a role in change of karyotypes. Arora et al. (1985) found 7.16–9.261 µ, 7.33–11.52 µ and 5.5–9.6 µ chromosome length in *C. tetragonoloba*, *C. serrata* and *C. senegalensis*, respectively and 44, 45 and 45 % total form (TF) values, respectively.

Polyploidy is a common technique to overcome the sterility of a hybrid species in plant breeding and is induced by treating seeds with the chemical, viz., colchicine. Vig (1963) induced polypoidy in clusterbean and observed a mixoploid plant resulting from aqueous colchicine treatment and obtained true autotetaploids. Bewal et al. (2009) induced autotetraploidy in *C. tetragonoloba* and found that the tetraploids plants has reduced plant height, length of rachis, length and breadth of leaflet and number of seeds/pod from the corresponding diploids plants along with considerable increase in internodal length, length and breadth of standard petal, width of wing petal and pod length. The tetraploid plants have normal meiotic behaviour and may lead to a good seed set.

2.4 Hybridization

Hybridization is the process of interbreeding between individuals of different species or genetically divergent individuals from the same species and is one of the important tools for creation of genetic variability in any crop (Harrison and Larson 2014). The manual hybridization in clusterbean is difficult and laborious due to highly self-pollination, small flower, its structures and flower drop. The earliness may be transferred from wild species (*C. serrata* and *C. senegalensis*) to cultivated species of clusterbean with the help of hybridization. Besides this, locally collected landraces can be used in breeding programme for incorporation of desirable traits in clusterbean. Multiple crossing, back cross, pedigree and bulk pedigree are the major varietal improvement programmes in a number of self-pollinated species for different inherited characters such as seed size, disease resistance, seed colour and duration of maturity (Knauft and Ozias-Akins 1995). But these methods are not found to be efficient for the improvement of quantitative traits like seed yield. Various physical and chemical mutagens have been reported for induction of variability in this crop.

2.4.1 Conventional Method of Hybridization

The interspecific hybridization between *C. tetragonoloba* and *C. serrata* through conventional methods was reported as complete failure probably due to the rejection of pollen by the stigma (Sandhu 1988). Other approaches like bud pollination, amputation of stigma and style, use of organic solvents also failed to overcome the stigmatic incompatibility barriers. Scope for interspecific hybridization is, therefore, limited in clusterbean. Ahlawat et al. (2013) studied three species of *Cyamopsis* to find out barriers of interspecific crosses between *C. tetragonoloba* × *C. serrata* and *C. tetragonoloba* × *C. senegalensis* for crop improvement. They reported interspecific hybridization between *C. tetragonoloba* × *C. serrata* with least (10.43 %) pod setting.

2.5 Heterosis and Combining Ability

Heterosis and combining ability have been used as an important breeding approach in crop improvement for selection of parents and utilization of hybrid vigour and is a good method to assess the nature of gene action involvement in the inheritance of character (Vasal 1998). The knowledge of gene action and combining ability analysis gives an insight for identifying the better combiners which may be hybridized to exploit heterosis for selection of better crosses in further breeding work and also elucidate the nature and magnitude of various types of gene effects involved in the expression of quantitative traits (Nigussie and Zelleke 2001).

Several workers found the existence of wide range of heterosis in clusterbean (Chaudhary et al. 1981) but heterosis breeding has its own limitations in this crop due to absence of stable source of male sterility. Different gene effects were observed in different crosses and it was noticed that additive, dominance and epistatic effects were operating for both endosperm and gum content in clusterbean. Heterosis and combining ability studies for seed yield and component characters in clusterbean have been reported by number of workers (Saini et al. 1990; Arora et al. 1998). Hooda et al. (1999) found high specific combining ability (SCA) effects for number of characters with the crosses like Durgajay × AG 111, HFG 516 × HFG 590 and AG 111 × HFG 516 and suggested that the genic interactions for best crosses were accountable to additive × additive or additive × dominance or dominance × dominance type of gene effects. The cross HFG 516 × HFG 590 was recorded with high heterosis, low inbreeding depression and high SCA effect for seed yield, component characters and disease resistance.

The extent of heterosis and combining ability in this crop gave substantial support for developing hybrids but its possibility is ruled out in the absence of techniques for economic production of large quantities of hybrid seed.

2.6 Mutation

Mutation is a sudden heritable change and is caused by alteration in the base gene sequence. It can be induced either spontaneously or artificially both in seed and vegetatively propagated crops (Bhosle and Kothekar 2010). The utilization of mutagenesis has been established as one of the best preference in the crops improvement (Dubinin 1961). Mutagenesis involves screening of genetic material and the selection of individual mutants having improved traits and their incorporation into breeding programmes. The efficient mutagenesis should be free from association with undesirable changes for creation of desirable changes. Mutations occur unexpectedly and sudden in nature and the frequency of such mutations is very low so breeder cannot rely on this for plant breeding. Therefore, artificial means using physical and chemical mutagens were discovered to induce mutations in the plant. A number of physical and chemical mutagens known for their mutagenesis competences have been used in different crops (Mullainathan et al. 2014). Physical mutagen includes seed treatment with X-ray and gamma rays, whereas chemical mutagen includes ethyl methane sulphonate (EMS), diethyl sulphonate (DES), sodium azide (SA), methyl methane sulphonate (MMS), nitroso guandine (NG), nitroso methyl urea (NMU), etc.

2.6.1 Natural Mutation

Naturally occurring genetic mutations may be caused by breaks, chimaeras, etc., and can change the appearance of the foliage, flowers, fruit or stem of any plant and may lead to abnormal plants (Tariq et al. 2008). A mutant of clusterbean having rosette-type inflorescence (raceme) with reduced fertility was reported in South Africa (Stafford 1989). The first report of the male sterility and partial male sterility in clusterbean was published from India (Mittal et al. 1968). Semi-sterile plants reported by Mittal et al. (1968) and Vig (1965) were shown to be caused by reciprocal translocations and produced fewer seeds/pod and more racemes/plant, grew taller and remained vegetative for longer time as compared to fertile plants. The practical utility of male sterility, partial male sterility and rosette raceme mutants has not been demonstrated in clusterbean (Arora and Pahuja 2008).

2.6.2 Mutations Induced Through Physical Factors

Physical mutation has been a successful tool in bringing improvement in self-pollinated crops and provides beneficial variation for practical plant breeding purpose (Tariq et al. 2008). It can be induced by irradiation with ionizing or non-ionizing rays. In the beginning, X-rays were used for this purpose but later gamma rays, viz., ^{60}Co and ^{137}Cs became more popular (Auerbach and Robson 1946). Gamma rays are known to influence plant growth and development by inducing cytological, genetical, biochemical, physiological and morphogenetic changes in cell and tissue (Gunckel and Sparrow 1961). The first successful attempt of mutation induced through physical mutagens in clusterbean was carried with gamma rays using ^{60}Co as source of radiation (Vig 1965). The effects of irradiation on morphology and cytology of clusterbean using gamma rays were studied (Lather and Chaudhary 1972) and decreased germination percentage, seedling survival and pollen fertility was recorded with increased dosage. The dose of 100–200 kR was proved to be lethal with no germination. By contrast, Chaudhary et al. (1973) observed increased yield, protein and gum contents in an M_2 population generated through irradiation with low doses of gamma rays.

The effects of irradiation ranging from 10 to 250 kR gamma rays on clusterbean for quantitative characters assessed in the M_2 generation (Chowdhury et al. 1975) showed that the number of branches and seed yield/plant were less in the M_2 generation than the control but the peduncle length and plant height were greater. Similarly, M_2 progenies of clusterbean obtained through gamma irradiation was recorded with increased plant height, number of clusters/plant, number of pods/cluster, number of pods/plant and seed yield/plant but the induced variation reduced the number of seeds/pod and pod length (Amrita and Jain 2003). The yield potential of M_3 progenies derived from certain high yielding M_2 progenies of the clusterbean variety exposed to various doses of gamma rays was recorded with more number of pods/plant and seed yield (Yadav et al. 2004). Patil and Rane (2015) studied the frequency and spectrum of chlorophyll mutations in clusterbean using different doses of gamma rays and reported increased frequency of chlorophyll mutations with increasing doses of gamma rays.

The effect of 0–60 kR doses of X-rays was studied on seeds of clusterbean variety Pusa Navbahar, the low doses were found to be beneficial, while high doses were inhibitory, which were generally less in the second generation (Rao and Rao 1982). Early flowering and determinate type are the remarkable mutants derived with gamma irradiation. An early flowering mutant with more number of pods was generated from the 10 kR treatment of X-rays. The determinate plants have reduced plant height, non-branching habit, synchronous and early maturity, increased cluster size and the main shoot either terminated into a leaf or an inflorescence (Singh et al. 1981).

Among physical mutagens, ultraviolet (UV) rays are non-ionizing and are effective in producing purine or pyrimidine dimers, resulting in point mutations. Studies suggest that UV rays can be effectively used to irradiate pollen in the late or early uninucleate stages (Toker et al. 2007). Lingakumar and Kalandaivelu (1998) subjected clusterbean variety Pusa Navbahar seedlings to continuous UV-B radiation for 18 h, post-irradiated with white light and UV-A enhanced fluorescent radiations and found that UV-B alone reduced plant growth, pigment content and photosynthetic activities. While, supplementation of UV-A promoted overall seedling growth and enhanced the synthesis of chlorophyll and carotenoids with a relatively high photosynthesis. Joshi et al. (2007) reported decline in the amount of photosynthetic pigments, O_2 evolution with induced UV-B radiation and modification in the absorption spectra of chloroplasts, whereas the combination of UV-A and UV-B irradiation partially reversed these changes.

Arora et al. (1997a) carried out the genetic analysis of seed mass following hybridization and irradiation in clusterbean and observed that the additive component of genetic variance for the inheritance of 100-seed weight was more important in irradiated generations (F_1M_1 and F_2M_2), whereas non-additive gene action was more important in case of non-irradiated populations (F_1 and F_2). Vig (1969) observed that the seed of clusterbean had initially higher germination with the dose of 10 kR gamma rays but in the later hours 30 kR treated material gave higher germination percentage. Radiation speeds up the growth of roots, but there is no correlation between rate of root elongation and size of dose, however, a direct correlation between the radiation dose and the rate of germination was observed (Vig 1969). The comparison of nature and magnitude of variability among unirradiated and irradiated populations of clusterbean revealed that the irradiated populations had higher seed weights, early flowering and maturity (Arora and Lodhi 1995). The physical mutagens often result in the larger scale deletion of DNA and changes in chromosome structure. Therefore, presently they have been shifted to chemical mutagens.

2.6.3 Mutations Induced Through Chemical Factors

As ionizing radiation results in high rates of chromosome aberrations and is accompanied with detrimental effects, it was obvious to look for alternate sources and as a result chemical mutagens have been discovered. The chemical-induced mutations have significant importance in evaluation and crop improvement and are considered as more viable mutants (Heslot et al. 1961). The most widely used chemical mutagens are alkylating agents, among these EMS being the most popular because of its effectiveness and ease of handling, especially its detoxification through hydrolysis for disposal (Hajra 1979). EMS usually creates point mutations in plants (Okagaki et al. 1991), whereas SA creates marginal mutagenic in different organisms (Arenaz et al. 1989). SA is inexpensive, relatively safe to handle and has no carcinogenic effect and has also been reported to induce high frequency of point mutation without detectable chromosomal aberrations (Nilan et al. 1973) and due to the efficiency of mutagenic effects of SA, no new chemical mutagens of widespread use in plant breeding have been discovered (Sikora et al. 2011).

The M_1 plants of clusterbean developed from EMS-treated seeds were chlorophyll deficient and showed profuse vegetative growth (Gohal et al. 1972). Similarly, mutants with changed leaf texture or shape, growth habit and pod size were obtained by treating seeds with EMS or hydrazine hydrate and unbranched and regular pod bearing mutants were obtained by treating the seeds with hydroxyl amine (Swamy and Hashim 1980). Some mutants showed changes in seed colour from normal violet to light grey or light brown. Rao et al. (1982) observed determinate and spreading variants in the M_1 and M_2 generations of Pusa Navbahar, when soaked in 200, 400 and 600 ppm kitazin and 1000, 2000 and 3000 ppm Saturn for 12 and 24 h. Velu et al. (2008) induced high frequency of viable mutations in M_2 generation of clusterbean by both EMS and SA. They further reported increased mutation frequency with increased dose of SA, while in case of EMS the mutation frequency decreased with increasing doses. Unbranched and regular pod bearing mutants (Swamy and Hashim 1980), chlorophyll deficient and mutants with profuse vegetative growth (Gohal et al. 1972) has been observed and isolated by treating the seeds with EMS and hydroxyl amine. Further, the seeds treated with EMS or hydrazine hydrate resulted into changed leaf structure, growth habit and pod size in the populations of the clusterbean, some of the mutants were recorded with extensive branching and late flowering and some were recorded with changes in seed colour (Swamy and Hashim 1980).

2.6.4 Mutations Induced Through Physical and Chemical Factors

The successfulness of mutagen depends on its effectiveness and efficiency. The effectiveness generally means the rate of point mutations relative to dose, whereas efficiency refers to the rate of point mutations relative to other biological effects induced by the mutagen and is considered a measure of damage (Konzak et al. 1965). Mutation induction using both physical and chemical mutagens suggest that the lower doses of irradiation coupled with chemical mutagens are popular for in vitro mutation (Medina et al. 2004). In clusterbean the mutagenic efficiency and effectiveness were decreased with an increase in the dose of gamma rays or concentration of EMS in alone or gamma rays followed by EMS in combination treatments. Gamma rays in alone induced more mutagenic efficiency due to their high penetrating power as compared to EMS in alone or gamma rays followed by EMS in combination treatments (Dube et al. 2011).

Velu et al. (2007a) estimated the mutagenic effectiveness and efficiency of gamma rays and EMS in the clusterbean and reported that the frequency and efficiency of mutation was more in EMS as compared to gamma rays. Their finding suggests that EMS induced more number of mutants effectively and efficiently than gamma rays. Further, Velu et al. (2007b) reported that increasing dose/concentration of gamma rays and EMS decreased the values of morphological and yield parameters in M_1 generation. Similarly Mahla et al. (2010) studies the effectiveness and efficiency of gamma rays and EMS in clusterbean and observed steady reduction in germination and subsequent survival of the treated population, seedling height and pollen fertility with increasing doses/concentration of mutagens. Babariya et al. (2008) treated the seeds of clusterbean with gamma rays and EMS and studied the effect of mutagenic treatments on the character association in M_2 generation. The results revealed that besides generating variability, mutagenic treatments can alter the mode of association between any two characters and selection for improved plant architecture.

The reduction of germination, survival, growth of seedlings, plant height, number of leaves/plant, number of branches/plant, number of pods/plant, number of clusters/plant, pod length, pod breadth, fresh and dry weight of matured plant were decreased with increasing doses and concentration of gamma rays and EMS in clusterbean, whereas days to first flowering increased with increasing doses and concentration of gamma rays and EMS (Velu et al. 2012). The mutation frequency was increased with the increase in the mutagen dose of gamma rays, whereas in case of EMS it decreased with the increase in doses. Yadava and Chowdhury (1974) studied the cytological effects of different doses of gamma rays and sodium nitrate and found that 100 and 150 kR doses of gamma rays were lethal and caused the highest abnormalities in different parts of the plant. High yielding mutants were induced by treating the clusterbean seeds with gamma rays and aqueous solutions of EMS and NMU either alone or in various combinations, the M_2 mutants thus obtained had long pods, increased number of pods and early maturity. These mutants also showed increased yields and gum contents (Singh and Aggarwal 1986). Basha and Rao (1988) studied combined mutagenic effects of gamma rays and SA in clusterbean varieties and found SA as the most efficient mutagen.

Mullainathan et al. (2014) reported higher percentage of mitotic aberrations with higher dosage of gamma radiation, EMS, SA and colchicine treated plants.

Along with broken metaphases, anaphasic laggards/bridges, clumping up of chromosomes, unequal anaphase and precocious movements of chromosomes during metaphases have also been observed in the plants treated with mutagenic agents. The mitotic aberrations are more in colchicine treated plants than the plants treated with gamma rays, EMS and SA.

Use of physical mutagens coupled with chemical mutagens has opened new outlook in the field of hybridization of this crop and there is requirement of more planned studies to obtain more desirable and fruitful mutagens of clusterbean. The lower doses of gamma rays (up to 50 kR) and lower concentrations of EMS (below 0.2 %) may be used in future breeding programme for inducing broad spectrum and high frequency of viable mutations in this crop.

2.7 Genotype × Environment Interaction and Stability

Genotype × environment interaction (G × E) is the change in the relative performance of a character of two or more genotypes studied in two or more environments and involves changes in rank order for genotypes between environments and changes in the absolute and relative magnitude of the genetic, environmental and phenotypic variances between environments (Haldane 1946). The studies of G × E and adaptation involve relative observation of genotypic responses in terms of yield under target environmental conditions (Allard 1960). A variety may achieve stability through individual or population buffering and the yield is demonstrated as their effects (Allard and Bradshaw (1964). Similarly, when the varieties are compared over a series of environments, their relative rankings generally differ and this causes difficulty in demonstrating the significant superiority of any variety (Eberhart and Russell 1966). Thus, to evaluate the stability of a genotype for their yield a precise knowledge of G × E interaction is of vital importance and phenotypically stable lines are of great importance for the crops like clusterbean.

Stable varieties/genotypes of clusterbean over the wide range of environmental conditions have been screened over the years by different workers (Pathak et al. 2010) and all the workers emphasized the importance of genotype over environment, the linear regression of the genotype over environmental index and the deviation from regression coefficient for determination of stability and adaptability of the genotype for seed yield and other important yield influencing traits in clusterbean. Chaudhary et al. (2005) evaluated clusterbean for seed yield and its components over three environments for stability analysis. They observed that genotype, environment and G × E interaction were significant for all the traits indicating presence of variability for genotypes environments and non-linear response of genotypes over the environments. D'almeida and Tikka (2003) carried out stability analysis for seed yield and quality traits, viz., seed size, protein and gum content and reported significant G × E interaction for seed yield, protein and gum content while it was non-significant for seed size. Linear and non-linear components of

G × E interaction were also significant for seed yield, protein and gum content, whereas both were non-significant for seed size.

Studies in clusterbean for green fodder yield (Paroda and Mehrotra 1976) and dry matter yield (Jhorar et al. 1980) indicated that different genotypes behave differently under varying environmental conditions. Henry and Mathur (2005) studied three different environments for different quantitative and qualitative characters and found that the gum content was better under late sown environment, whereas protein content exhibited higher value under the crop sown with onset of monsoon. In the same study they found that the high gum content and low value of protein content was exhibited by early maturing genotypes. Jain and Patel (2012) reported stable varieties, viz., GAUG-0309, GAUG-0416, GAUG-0513 and GAUG-0522 for earliness; GAUG-0416, GAUG-0308, GAUG-0004 and GAUG-0309 for plant height; GAUG-0411 for pods/plant and GAUG- 0309 and GAUG-0411 for seed yield and suitable for cultivation in North Gujarat.

2.8 Correlation and Path Analysis

Selection of a variety is mainly based on phenotypic characters in the breeding programme but the response to selection depends on many factors including the information on association of characters, direct and indirect effects contributed by each character (Mohammadi et al. 2003). Correlation and path analysis establish the magnitude and direct and indirect effects of relationship between yield and its components. Correlation estimates the degree of association between the variables whereas, path coefficient analysis provides an indication that which variables exert an influence on other variables (Akanda and Mundit 1996). The correlation coefficient between the predictor and response variable is partitioned using path coefficient analysis, which provides a method of separating direct and indirect effects and measuring the relative importance of the causal factors involved to develop selection criteria for complex traits in several crop species (Milligan et al. 1990).

The correlation studies in clusterbean suggest that a plant having few branches, more number of clusters and pods, bolder seeds, long peduncles is expected to have higher seed yields (Vijay 1988). Shekhawat and Singhania (2005) observed that pods/plant, branches/plant, clusters/plant, pods/cluster, 100-seed weight and plant height had direct and positive effect on seed yield. Seed yield was significantly and positively correlated with plant height, seeds/pod, pods/plant, primary and secondary branches/plant. Pods/plant, seeds/pod and 100-seed weight had the maximum positive effect on seed yield/plant. Also, plant height, seeds/pod, gum content, primary and secondary branches had sizeable indirect effect via pods/plant. 100-seed weight had a positive direct effect on yield and positive indirect effects via clusters/plant. It is desirable to improve both seed weight and clusters/plant for better seed yield (Arora et al. 1997b). Thus, pods/plant, branch numbers, plant height and 100- seed weight may be considered as effective parameters of selection to increase seed yield in clusterbean

(Pathak et al. 2011b). Studies revealed that grain yield and gum content has positive correlation while seed weight has negative correlation with gum content, whereas endosperm always has a positive correlation with gum content. Similarly, seed yield and gum content were positively correlated with height, branch number and pod number whereas it had negative association with pod length and 100-seed weight, similarly endosperm had negative association with seed size and pod length (Pathak et al. 2011b).

Seed yield/plot was significantly and positively correlated with biological yield/plot, number of clusters and pods/plant and plant height. Stafford and Seiler (1986) recorded high and positive correlations between seeds/pod and seed yield and pods/plant and seed yield. The correlation of 100-seed weight and seed yield was low and positive. Grain yield/plant was found to be positively and significantly associated with all characters except gum content and pod length. Number of seeds/pod, number of pods/plant and pod length was the most important component characters which directly contributed to seed yield (Patel and Chaudhari 2001). Vir and Singh (2015) reported positive and significant correlations of number of seeds/pod, number of pods/plant, number of pods/cluster, number of clusters/plant, days to 50 % flowering and days to maturity with seed yield/plant. The positive correlation between number of pods/plant and seed yield was mainly due to positive direct effect of pods/plant (Singh et al. 2004). Weixin et al. (2009) reported negative association of quality related characters to seed yield. There are positive and significant correlation of seed yield with dry pod yield, number of pods/cluster, 100–seed weight, number of clusters/plant, branches/plant, seed recovery, germination, number of seeds/pod and dry biomass/plant. Whereas negative and significant correlation was recorded in days to flower initiation, plant height, days to maturity and days to 50 % flowering. Plant height, number of seeds/pod, days to 50 % flowering, number of clusters/plant, dry pod yield/plant and dry biomass/plant had direct positive effects on seed yield. Kumar and Ram (2015) suggested that the number of clusters/plant, number of pods/plant, pod yield/plant, plant height and days to maturity are the important traits for selection to yield improvement in clusterbean.

Anandhi and Oommen (2010) reported positive association of number of pods/ plant, number of seeds/pod, pod weight and number of pod clusters/plant with vegetable pod yield and suggested that selection based on number of pods/plant, number of seeds/pod, pod weight and number of pod clusters/plant may bring out desired improvement towards enhancing the vegetable pod yield in clusterbean besides this selection of dwarf and early flowering genotypes would result in better yielding types. Girish et al. (2012) reported positive association of vegetable pod yield with pod breadth whereas Shabarish and Dharmatti (2014) observed positive and significant correlation of pods/plant with vegetable pod yield/plant in clusterbean. Malaghan et al. (2014) studied correlation among the vegetable pod yield components and their direct and indirect effects on the vegetable pod yield of clusterbean. The study revealed that the pod yield/plant was significantly and positively associated with pod length, pod breadth, pods weight and pods/plant. The pods/plant, pod length, pod weight and pod breadth had positive direct effect on vegetable pod yield/plant. They suggested that the parameters, viz., pods/plant, pod length, ten fresh pod weight and pod breadth may be the potential traits for selection of higher yielding vegetable pod genotypes.

Genetic path coefficient analyses showed that pods/plant and 100-seed weight were important factors in determining seed yield whereas, seeds/pod was least important indicating that most of the yield components contributes via number of clusters/plant and had maximum direct effect as well as positive correlations with yield (Arora and Gupta 1981) and revealed positive association of number of seeds/pod, number of branches/plant and number of seeds/pod with the seed yield. Number of clusters on branches had highest direct effect on seed yield/plant followed by the clusters on main stem and pods on branches. The total number of branches on the plant, days to maturity and total number of pods contributed towards grain yield/plant via number of clusters on branches followed by the number of pods on the branches. Number of clusters on the branches in turn contributed to the yield directly as well as via number of pods on the branches and total number of pod. Thus, number of clusters and pods on the branches are the most cordial component of seed yield in clusterbean.

2.9 Genetic Resources

Development of newer materials in the form of varieties is an essential requirement for increased and sustainable production of the crop. The low productive clusterbean varieties and the varieties inclined to one or the other diseases and pests, needs to be replaced with the newer ones. Increase of genetic resources through germplasm collection both from indigenous and exotic sources is the most important criterion for any crop breeding and improvement programme.

National Bureau of Plant Genetics Resources, New Delhi have collected about 5000 accessions of clusterbean from dry habitats of northern India including two wild species viz., *C. serrata* and *C. senegalensis*. 4878 accessions with indigenous origin have also been conserved in medium term storages and 3714 accessions have been put for *ex situ* conservation (Mishra et al. 2009). Effective evaluation of more than 375 accessions against important diseases have resulted in promising resistance lines against bacterial leaf blight (GAUG-9406, GG-1, RGC-1027), Alternaria leaf blight (GAUG-9406, GAUG-9005, GG-1, GAUG-9003) and root rot (GG-1, HGS-844, GAUG-9406) (Kumar 2008). Certain lines of clusterbean, viz, Sona, Suvidha, IC-09229/P3, Naveen, PLG-85 and RGC-471 for seed type, and others like, Pusa Mausmi, Pusa Sadabahar, Pusa Navbahar, IC-11388, PLG-850 and Sharad Bahar were released as promising varieties for vegetable purposes.

References

Ahlawat A, Pahuja SK, Dhingra HR (2013) Overcoming interspecific hybridization barriers in *Cyamopsis* species. Int J Biotech Bioengin Res 4(3):181–190

Akanda SI, Mundit CC (1996) Path coefficient analysis of the effects of stripe rust and cultivar mixtures on yield and yield components of winter wheat. Theor Appl Genet 92:666–672

Allard RW (1960) Principles of plant breeding. Wiley, New York

Allard RW, Bradshaw AD (1964) Implications of genotype environment interactions in applied plant breeding. Crop Sci 4:503–508

Amrita KR, Jain UK (2003) Induction of variability through gamma irradiation in guar (*Cyamopsis tetragonoloba* L. Taub.). Progressive Agric 3:121–122

Anandhi K, Oommen SK (2007) Variability and heritability of yield and related characters in clusterbean (*Cyamopsis tetragonoloba* (L.) Taub.). Legume Res 30(4):287–289

Anandhi K, Oommen SK (2010) Correlation studies in clusterbean. Legume Res 33(3):227–228

Arenaz P, Hallberg L, Mancillas F et al (1989) Sodium azide mutagenesis in mammals; inability of mammalian cells to convert azide to mutagenic intermediate. Mutat Res 277:63–67

Arora SS, Gupta BS (1981) Genetic variability correlation and path coefficient analysis in guar (*Cyamopsis tetragonoloba* (L.) Taub.). Curr Agric 5(3–4):116–122

Arora RN, Lodhi GP (1995) Genetic variability in guar following hybridization and irradiation. Indian J Plant Genet Res 8(2):201–208

Arora RN, Pahuja SK (2008) Mutagenesis in Guar [*Cyamopsis tetragonoloba* (L.) Taub.]. Plant Mutat Rep 2(1):7–9

Arora RN, Sareen PK, Saini ML et al (1985) Karyotype analysis in three species of genus *Cyamopsis*. Indian J Genet 45(2):302–309

Arora RN, Lodhi GP, Pahuja SK et al (1997a) Genetic analysis of seed mass following hybridization and irradiation in clusterbean (*Cyamopsis tetragonoloba* L. Taub.). Ann Biol 13:59–65

Arora RN, Lodhi GP, Singh JV (1997b) Correlation and path analysis in F1, F1M1, F2 and F2M2 generations in clusterbean. Forage Res 23(1&2):31–38

Arora RN, Lodhi GP, Singh JV (1998) Variety cross diallel analysis for grain yield and its components in clusterbean (*Cyamopsis tetragonoloba* (L.) Taub.). Forage Res 24(1):1–6

Arora RN, Lodhi GP, Singh JV et al (1999) Genetic analysis of grain yield and its component characters in F2 population of guar. Ann Biol 15(1):45–49

Auerbach C, Robson JM (1946) The chemical production of mutations. Nature 157:302

Babariya HM, Vaddoria MA, Mehta DR et al (2008) Effect of mutagens on characters association in clusterbean (*Cyamopsis tetragonoloba* L Taub). Int J Biosci Reporter 6(1):135–140

Basha SK, Rao PG (1988) Gamma ray and sodium azide induced heterophylly of bhindi and clusterbean. J Neucl Agric Biol 17:133–136

Bewal S, Purohit J, Kumar A et al (2009) Cytogenetical investigations in colchicine-induced tetraploids of *Cyamopsis tetragonoloba* L. Czech J Genet Plant Breed 45(4):143–154

Bharodia PS, Zaveri PP, Kher HR et al (1993) GAUG 34 a high yielding variety of clusterbean. Indian Farm 43(9):31–33

Bhosle SS, Kothekar VS (2010) Mutagenic efficiency and effectiveness in clusterbean (*Cyamopsis tetragonoloba* (L.)Taub.). J Phytol 2(6):21–27

Burton GW, De Vane EH (1953) Estimating genetic variability in tall fescue (Festucarundinacea) from replicated clonal material. Agron J 45:478–481

Chaudhary MS, Ram H, Hooda RS et al (1973) Effect of gamma irradiation on yield and quality of guar (*Cyamopsis tetragonoloba* L. Taub). Ann Arid Zone 12:19–22

Chaudhary BS, Lodhi GP, Arora ND (1981) Heterosis for grain yield and quality characters in clusterbean. Indian J Agric Sci 51(9):638–642

Chaudhary SPS, Chaudhary AK, Shekhawat SS et al (2003) Quantitative genetic analysis in some genotypes of clusterbean. In: Henry A, Kumar D, Singh NB (eds) Advances in arid legumes research. Scientific Publishers (India), Jodhpur, pp 9–13

Chaudhary SPS, Singh RV, Singh NP et al (2005) Stability for seed yield influencing traits in clusterbean (*Cyamopsis tetragonoloba* (L.) Taub.). J Arid Legumes 2(1):86–90

Chowdhury RK, Chawdhury JB, Singh RK (1975) Induced polygenic variability in clusterbean. Crop Improv 2:17–24

D'almeida A, Tikka SBS (2003) Stability of yield and quality of traits in guar. In: Henry A, Kumar D, Singh NB (eds) Advances in arid legumes research. Scientific Publishers (India), Jodhpur, pp 367–371

Dabas BS (1975) Studies on inheritance of quantitative characters and gum content in guar (*Cyamopsis tetragonoloba* L. Taub). Ph.D. Thesis (unpublished), IARI, New Delhi

Dube KG, Bajaj AS, Gawande AM (2011) Mutagenic efficiency and effectiveness of gamma rays and EMS in *Cyamopsis tetragonoloba* (L.) var. *Sharada*. Asiatic J Biotech Resour 2(4):436–440

Dubinin NP (1961) Problems of radiation genetics. Oliver and Boyed, London

Eberhart SA, Russell WW (1966) Stability parameters for comparing varieties. Crop Sci 6:36–40

Girish MH, Gasti VD, Mastiholi AB et al (2012) Correlation and path analysis for growth, pod yield, seed yield and quality characters in clusterbean (*Cyamopsis tetragonoloba* (L.) Taub.). Karnataka J Agric Sci 25(4):498–502

Girish MH, Gasti VD, Shantappa T et al (2013) Genetic variability studies in clusterbean [*Cyamopsis tetragonoloba* (L.) Taub.]. Karnataka J Agric Sci 26(3):442–443

Gohal MS, Kalia HR, Dhillon HS et al (1972) Effect of ethyl methane sulphonate on the mutation spectrum in guar. Indian J Hered 2:51–54

Gunckel JE, Sparrow AH (1961) Ionizing radiation: biochemical, physiological and morphological aspects of their effects on plants. In: Ruhland W (ed) Encyclopaedia of plant physiology, vol XVI. Springer, Berlin, pp 555–611

Hajra NG (1979) Induction of mutations by chemical mutagens in tall indica rice. Indian Agric 23:67–72

Haldane JBS (1946) The interaction of nature and nurture. Ann Eugen (London) 13:197–205

Harrison RG, Larson EL (2014) Hybridization, introgression, and the nature of species boundaries. J Hered 105:795–809

Henry A, Mathur BK (2005) Genetic diversity and performance of clusterbean varieties for quality and quantitative characters in arid region. J Arid Legumes 2(1):145–148

Henry A, Daulay HS, Bhati TK (1992) Maru Guar promising clusterbean for arid region. Indian Farm 42(6):24–25

Heslot HR, Ferary R, Levy R et al (1961) Effect of ionizing radiation of seeds. International Atomic Energy Agency, Viana, pp 248–249

Hooda JS, Saini ML, Rai L (1999) Heterosis and combining ability studies in clusterbean (*Cyamopsis tetragonoloba* (L.) Taub.). Forage Res 25(3):191–193

Jahan B, Vahidy AA, Ali SI (1994) Chromosome number in some taxa of Fabaceae mostly native to Pakistan. Ann Missouri Bot Gard 81:10–15

Jain SK, Patel PR (2012) Stability analysis for seed yield and their component traits in breeding lines of guar (*Cyamopsis tetragonoloba* L.). Legume Res 35(4):327–331

Jhorar BS, Saini ML, Solanki KR (1980) Phenotypic stability of dry matter yield in clusterbean. Haryana Agric Univ J Res 10:1–4

Joshi PN, Ramaswamy NK, Iyer RK et al (2007) Partial protection of photosynthetic apparatus from UV-B-induced damage by UV-A radiation. Environ Exp Bot 59:166–172

Kapoor R, Bajaj RK (2014) Genetic variability and association studies in guar (*Cyamopsis tetragonoloba* (L.) Taub.) for green fodder yield and quality traits. J Res Punjab Agric Univ 51(2):109–113

Knauft DA, Ozias-Akins P (1995) Recent methodologies for germplasm enhancement and breeding. In: Pattee HE, Stalker HT (eds) Advances in peanut science. American Peanut Research and Education Society, Stillwater, pp 54–94

Konzak CF, Nilan RA, Wagner J et al (1965) Efficient chemical mutagenesis. Rad Bot 5:49–70

Kumar D (2008) Arid legumes-an introduction. In: Kumar D, Henry A (eds) A souvenir, 3rd national symposium on arid legumes, June 28–30, 2008. Indian Arid Legumes Society, CAZRI, Jodhpur, pp 25–44

Kumar V, Ram RB (2015) Genetic variability, correlation and path analysis for yield and yield attributing traits in cluster bean [*Cyamopsis tetragonoloba* (L.) Taub.] genotypes. Int J Pure Appl Biosci 3(1):143–149

Lather BPS, Chowdhury JB (1972) Studies on irradiated guar. Nucleus 15:16–22

Lingakumar K, Kalandaivelu G (1998) Differential responses of growth and photosynthesis in Cyamopsis tetragonoloba L. grown under ultraviolet-B and supplemental long-wavelength raditations. Photosynthetica 35:335–343

Mahla HR, Kumar D, Shekhawat A (2010) Effectiveness and efficiency of mutagens and induced variability in clusterbean (*Cyamopsis tetragonoloba*). Indian J Agri Sci 80(12):1033–1037

Malaghan SN, Madalageri MB, Kotikal YK (2014) Correlation and path analysis in cluster bean [*Cyamopsis tetragonoloba* (L.) Taub.] for vegetable pod yield and its component characters. Bioscan 9(3):1583–1586

Maniee M, Kahrizi D, Mohammadi R (2009) Genetic variability of some morpho-physiological traits in Durum wheat (*Triticum turgidum* var. durum). J Appl Sci 9(7):1383–1387

Medina FIS, Amano E, Tano S (2004) FNCA mutation breeding manual. Forum for Nuclear Cooperation in Asia. Available from: URL: http://www.fnca.mext.go.jp/english/mb/mbm/e_mbm.html. Accessed 25 May 2014)

Milligan SB, Gravois KA, Bischoff KP et al (1990) Crop effects on genetic relationships among sugarcane traits. Crop Sci 30:927–931

Mishra SK, Singh N, Sharma SK (2009) Status and utilization of genetic resources of arid legumes in India. In: Kumar D, Henry A (eds) Perspective research activities of arid legumes in India. Indian Arid Legumes Society, CAZRI, Jodhpur, pp 23–30

Mittal SP, Dabas BS, Thomas TA (1968) Male sterility in guar (*Cyamopsis tetragonoloba* (L.) Taub.). Current Sci 37:357

Mohammadi SA, Prasanna BM, Singh NN (2003) Sequential path model for determining interrelationships among grain yield and related characters in maize. Crop Sci 43:1690–1697

Mullainathan L, Aruldoss T, Velu S (2014) Cytological studies in cluster bean by the application of physical and chemical mutagens (*Cyamopsis tetragonoloba* L.). Int Lett Nat Sci 16:35–40

Nigussie M, Zelleke H (2001) Heterosis and combining ability in a diallel among eight elite maize populations. Afr Crop Sci J 9:471–479

Nilan RA, Siders EG, Kleinhofs A et al (1973) Azide—a potent mutagen. Mutat Res 17:142–144

Ogwu MC, Osawaru ME, Ahana CM (2014) Challenges in conserving and utilizing plant genetic resources. Int J Genet Mol Biol 6(2):16–22

Okagaki RJ, Neuffer MG, Wessler SR (1991) A deletion common to two independently derived waxy mutations of maize. Genet 127:425–431

Paroda RS, Mehrotra N (1976) An investigation of genotypes/environment interactions for green fodder yield in clusterbean (*Cyamopsis tetragonoloba* (L.) Taub). Genet Agrar 30:191–200

Patel BV, Chaudhari FP (2001) Component analysis of yield in clusterbean. Forage Res 27(2):123–126

Pathak R, Singh SK, Singh M et al (2010) Performance and stability of *Cyamopsis tetragonoloba* (L.) Taub genotypes under rainfed conditions. Indian J Dryland Agric Res Dev 25(2):87–92

Pathak R, Singh M, Henry A (2011a) Stability, correlation and path analysis for seed yield and yield attributing traits in clusterbean. Indian J Agric Sci 81(4):309–313

Pathak R, Singh M, Henry A (2011b) Genetic diversity and interrelationship among clusterbean genotypes for qualitative traits. Indian J Agric Sci 81(5):402–406

Pathak R, Singh M, Singh SK et al (2011c) Genetic variation in morphological traits and their interrelationships with gum content and yield in clusterbean. Ann Arid Zone 50(1):77–79

Patil CG (2004) Nuclear DNA amount variation in *Cyamopsis* D.C. (Fabaceae). Cytologia 69:59–62

Patil BM, Rane GM (2015) Gamma radiation induced chlorophyll mutations in clusterbean
    (*Cyamopsis tetragonoloba* (L.)Taub) Var. NCB-12. Int J Allied Pract Res Rev 2(2):75–85
Purohit J, Kumar A, Hynniewta M et al (2011) Karyomorphological studies in guar (*Cyamopsis
    tetragonoloba* (Linn.) Taub.)—An important gum yielding plant of Rajasthan, India.
    Cytologia 76(2):163–169
Raghu Prakash KR, Prasanthi L, Reddy SM (2008) Genetic variability studies for seed yield,
    physiological and quality attributes in guar (*Cyamopsis tetragonoloba* (L.) Taub). Asian
    Australasian J Plant Sci Biotech 2(1):36–38
Rai PS, Dharmatti PR, Shashidhar TR et al (2012) Genetic variability studies in clusterbean
    [*Cyamopsis tetragonoloba* (L.) Taub]. Karnataka J Agric Sci 25(1):108–111
Rao S, Rao D (1982) Studies on the effect of X-irradiation on *Cyamopsis tetragonoloba* (L.)
    Taub. Proc Indian Natl Sci Acad Part B Biol Sci 48:410–415
Rao SRM, Murthy PK, Rao D (1982) Note on determinate and spreading variants in clusterbean.
    Curr Sci 51(19):945–956
Saharan MS, Saharan GS, Singh JV (2004) Inheritance of branching behaviour in clusterbean
    [*Cyamopsis tetragonoloba* (L.) Taub.]. Forage Res 30(2):107–108
Saini ML, Arora RN, Paroda RS (1981) Morphology of three species of genus *Cyamopsis*. Guar
    Newsl 2:7–11
Saini ML, Singh JV, Jhorar BS et al (1990) Guar. Agric Sci Digest 10:113–116
Sandhu HS (1988) Interspecific hybridization studies in genus *Cyamopsis*. Ph.D. Thesis
    (Unpublished), CCS HAU, Hisar
Shabarish PR, Dharmatti PR (2014) Correlation and path analysis for cluster bean vegetable pod
    yield. Bioscan 9(2):811–814
Shekhawat SS, Singhania DL (2005) Correlation and path analysis in clusterbean. Forage Res
    30(4):196–199
Sikora P, Chawade A, Larsson M et al (2011) Mutagenesis as a tool in plant genetics, functional
    genomics, and breeding. Int J Plant Genom 2011:1–13. doi:10.1155/2011/314829
Singh VP, Aggarwal S (1986) Induced high yielding mutants in clusterbean. Indian J Agric Sci
    56:695–700
Singh VP, Yadav RK, Chowdhury RK (1981) Note on a determinate mutant of clusterbean.
    Indian J Agric Sci 51(9):682–683
Singh JV, Lodhi GP, Saini ML (1990a) Inheritance of leaf hairiness and leaf margin in cluster-
    bean (*Cyamopsis tetragonoloba* (L.) Taub). Ann Arid Zone 19(1):55–57
Singh JV, Saini ML, Lodhi GP et al (1990b) Genetics of gum content in guar (*Cyamopsis
    tetragonoloba* (L.) Taub). Forage Res 16(1):42–45
Singh JV, Kumar Y, Sharma S (2004) Interrelation between yield and its components in cluster-
    bean (*Cyamopsis tetragonoloba* (L.)Taub.). Forage Res 30(1):60–61
Singh M, Singh TP, Sharma SK et al (2010) Influence of cropping system on combining abil-
    ity and gene action for grain yield and its components in blackgram (*Vigna mungo*). Indian
    J Agric Sci 80(10):853–857
Singh P, Singh AK, Sharma M et al (2014) Genetic divergence study in improved bread wheat
    varieties (*Triticum aestivum*). Afr J Agric Res 9(4):507–512
Stafford RE (1989) Inheritance of partial male sterility in guar. Plant Breed 103:43–46
Stafford RE, Barker GL (1989) Heritability and interrelationships of pod length and seed weight
    in guar. Plant Breed 103:47–53
Stafford RE, Seiler GJ (1986) Path coefficient analyses of yield components in guar. Field Crops
    Res 14:171–179
Swamy LN, Hashim M (1980) Experimental mutagenesis in guar: some induced viable muta-
    tions of systematic interest. J Cytol Genet 15:61–63
Tariq MS, Javed IM, Haq MA et al (2008) Induced genetic variability in chickpea (*Cicer
    arietinum* L.) II Comparative mutagenic effectiveness and efficiency of physical and chemi-
    cal mutagens. Pak J Bot 40(2):605–613
Toker C, Yadav SS, Solanki IS (2007) Mutation breeding. In: Yadav SS, McNeil DL, Stevenson
    PC (eds) Lentil. Springer, Netherlands, pp 209–224

Vasal SK (1998) Hybrid maize technology: challenges and expanding possibilities for research in the next century. In: Vasal SK, Gonzalez CF, Xingming F (eds) Proceedings of the 7th Asian Reg. Maize Workshop, Los Banos, Philippines, February 23–27, pp 58–62

Velu S, Mullainathan L, Arulbalachandran D et al (2007a) Effectiveness and efficiency of gamma rays and EMS on clusterbean (*Cyamopsis tetragonoloba* (L.) Taub). Crop Res 34:249–251

Velu S, Mullainathan L, Arulbalachandran D et al (2007b) Effect of physical and chemical mutagen in clusterbean (*Cyamopsis tetragonoloba* (L.) Taub) in M_1 generation. Crop Res 34:252–254

Velu S, Mullainathan L, Arulbalachandran D et al (2008) Frequency and spectrum of morphological mutants in M_2 generation of clusterbean (*Cyamopsis tetragonoloba* (L.) Taub). Legume Res 31(3):188–191

Velu S, Mullainathan L, Arulbalachandran D (2012) Induced morphological variations in clusterbean (*Cyamopsis tetragonoloba* (L.) Taub). Int J Curr Trends Res 1(1):48–55

Vig BK (1963) Frequency of trivalents in autotetraploid guar. Curr Sci 32(8):375–376

Vig BK (1965) Effect of a reciprocal translocation on cytomorphology of guar. Sci Cult 31:531–533

Vig BK (1969) Studies with Co^{60} radiated guar (*Cyamopsis tetragonoloba* (L.) Taub). Ohio J Sci 69:18

Vijay OP (1988) Genetic variability, correlations and path analysis in clusterbean (*Cyamopsis tetragonoloba* (L.) Taub). Indian J Hort 45(1–2):127–131

Vir O, Singh AK (2015) Variability and correlation analysis in the germplasm of clusterbean [*Cyamopsis tetragonoloba* (L.) Taub.] in hyper hot arid climate of Western India. Legume Res 38(1):37–42

Weixin L, Anfu H, Peffley EB et al (2009) The inheritance and variation of gum content in guar [*Cyamopsis tetragonoloba* (L.) Taub.]. Agric Sci China 8(12):1517–1522

Yadav SL, Singh VV, Ramkrishna K (2004) Evaluation of promising M_3 progenies in guar (*Cyamopsis tetragonoloba* (L.) Taub.). Indian J Genet Plant Breed 64:75–76

Yadava JS, Chowdhury JB (1974) Cytological effects of physical and chemical mutagens on guar (*Cyamopsis tetragonoloba* (L.) Taub.). Haryana Agric Univ J Res 5:82–84

Chapter 3
Clusterbean Gum and By-Product

Abstract Clusterbean gum is a natural vegetable gum with a number of features, i.e., high natural viscosity due to its molecular structure, cold water solubility, rapidly hydration, freeze-thaw stability, etc. It is white to yellowish white in colour, odourless and is available in different viscosities and different granulometrics. It dissolves in polar solvent on dispersion and forms strong hydrogen bonds, while in nonpolar solvents it forms weak bonds. Its natural properties depend on its behaviour in an aqueous medium. The large number of patents has been registered throughout the world on various uses of clusterbean products in different applications. The structure, properties and estimation of gum along with its application and uses are discussed in detail in this chapter.

3.1 Introduction

The gums are sticky substance obtained from various plant species and often thought as resin, rubber, latex, etc. It was used by the ancient Egyptians for embalming the dead bodies and binding the mummies with strips of clothing. Likewise, gums have been used as food and for medicinal purpose by many civilizations. Clusterbean, xanthan, arabic, carrageenans, alginates, pectin, and various cellulose derivatives have been used as hydrocolloids (Guarda et al. 2004) for various purposes. Extensive surveys have been taken at national and international levels to search legume seeds bearing galactomannan. It includes guar gum, locust bean gum, tara gum, fenugreek gum, etc. that differ from one another in terms of their galactose: mannose (G:M) ratios (Garti and Leser 2001) and the pattern (ordered, random or as blocks) of galactose distribution along the molecular backbone (Dass et al. 2000). As a result, the number of legume plants seed bearing galactomannan polysaccharides has been identified, but only a few of these are being commercially used to produce seed gums. The classic characteristics for commercial production of seed galactomannan are

© Springer Science+Business Media Singapore 2015
R. Pathak, *Clusterbean: Physiology, Genetics and Cultivation*,
DOI 10.1007/978-981-287-907-3_3

1. It should be based on annual herb rather than perennial tree.
2. Its cultivation should be easy and can be increased by putting more land under cultivation when need arises.

Among various leguminous plants and trees, clusterbean seed is one of the major sources of galactomannan gum. The galactomannan gum from the clusterbean seed was developed as an industrial commodity as an exigency of World War II in year 1941. After World War II, there was major shortage of locust bean gum (LBG); due to occupation of most of the LBG-producing Mediterranean region by the Axis countries, adversely affected the textile and paper industries. Clusterbean gum was found as the most suitable substitute for scarce LBG at that time. Now it has got a boost after introduction of its anionic, cationic and hydrophobic non-ionic derivatives having numerous industrial applications.

Clusterbean gum is classified under mucilage and thickener group in harmonized system of classification in international trade. India is the largest exporter of clusterbean gum and mucilage with 38 % share of world mucilage and thickeners trade, followed by Spain (14 %), USA (9 %), Italy (7 %), Pakistan (5 %) and Germany (4 %). The trends of clusterbean export and use indicate that further intensive development efforts are likely to be continued in the years ahead (Sharma and Gummagolmath 2012). The global consumption of clusterbean gum and its derivatives was about 30 % in mid-seventies which has reached at about 55 % (Anonymous 2015). USA and Europe are the two key markets, which accounts for about 80 % of world imports of gum and splits of clusterbean. USA alone accounts for about 37 % of clusterbean gum exports from India followed by UK (16 %) and Germany (15 %). Other important countries are Hong Kong, Italy, Netherlands and Japan (Rana 2013). The export of clusterbean split is increasing which is an indication that the consumption of clusterbean derivatives has been increased.

3.2 Structural Feature of Galactomannan Gums

The chemical structure of clusterbean gum was established by the classical work of Whistler et al. (1950) and is based on the standard method of polysaccharide structure determination methods like determination of monosaccharide composition, methylation hydrolysis study, partial enzyme hydrolysis to determine the mode of linkage of various sugar components, molecular weight determination and periodate oxidation for determination of the extent of branching. Earlier, it was believed that the side groups of galactose and manose were substituted at regular intervals along the mannan backbone (Whistler and Hymowitz 1979) but the various studies, viz., enzyme degradation (McCleary 1979), spectroscopic methods (Grasdalen and Painter 1980) and computer simulation (McCleary et al. 1985) indicate random distribution of galactose side groups in clusterbean.

The legume seed galactomannan has a common structural feature and consists of a linear, polymer backbone of β, 1 → 4 linked mannopyranose which is randomly substituted by single α, 1 → 6 linked galactose grafts. Galactomannans are characterized to have variable range of molecular weight and the G:M ratio. Viscosity and other functional properties for the determination of application of galactomannan depend upon these characteristics.

Clusterbean gum forms a rod-like polymeric structure with a mannose backbone linked to galactose side chains, which randomly placed on mannose backbone with an average ratio of 1:2 galactose to mannose (Mudgil et al. 2014). Galactomannans of clusterbean gum differed in the ratio of galactose to mannose, viz., cold water-soluble fraction 1:1.3, while hot water-soluble fraction 1:1.7 water-insoluble ratio 1:7 and original sample 1:2, that reflect a wide spread of sugar compositions in and ensures polydispersity in molecular weight (Hui and Neukom 1964). The clusterbean gum contains 34.6 % D-galactopyranosyl units and 63.4 % D-mannopyranosyl units and consists of a linear chain of D-mannose units linked together by β (1 → 4) glycosidic linkage (Dodi et al. 2011). This galactose side stubs are attached as single side chain to the mannan backbone by α, (1 → 6) glycosidic linkage (Fig. 3.1). Each D-mannose unit bears a galactose unit on average basis. Both the sugars have hydroxyl groups in *cis*-position and are responsible for cross-linking property of the gum. McCleary et al. (1984) found that the D-galactosyl residues are arranged mainly in pairs of triplets and galactomannans from seeds of different clusterbean varieties are essentially identical.

The statistical dimensions of clusterbean gum from light scattering data were found to be 186.2 nm and the radius of gyration was 72.06 nm (Chakravorty et al. 1979). Winter et al. (1984) reported the three-dimensional structure of gum and found that the main chain of sugar molecules is grouped into a lamellar fashion and provides a space of variable dimensions for accommodating galactose and water molecules. Robinson et al. (1982) applied molecular theories of viscosities of dilute and more concentrated solutions of random polymer of clusterbean gum and revealed that the galactomannan chain is a coil-like molecule. Temperature plays an important role in the gelling behaviour of clusterbean gum and the gelling property is reduced with increase in temperature (Doyle et al. 2006). The configuration of clusterbean gum changes with the variation of temperature, and at low temperatures, it has a disc-like configuration whereas at higher temperature flexible linear configuration was observed (Relan et al. 1991). An increase in temperature causes the water

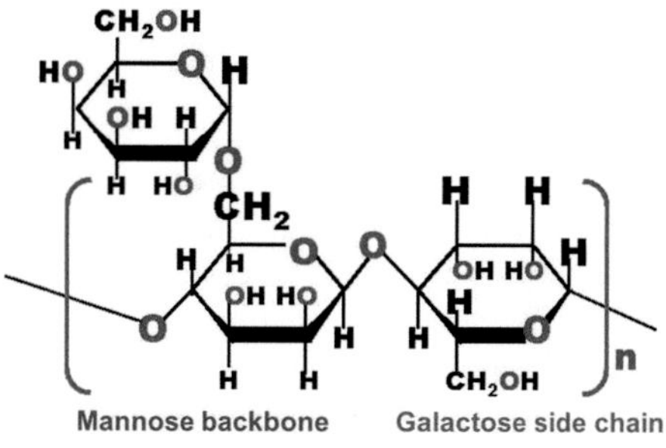

Fig. 3.1 Structure of clusterbean gum

Mannose backbone   Galactose side chain

molecules to lose their ordering around the molecule, and thus the conformation of clusterbean gum is disturbed resulting in reduced viscosity (Srichamroen 2007).

Clusterbean gum has the highest molecular weights as compared to other naturally occurring water-soluble polymers (Casas et al. 2000). The assessment of average molecular weight of clusterbean gum also varies on the methodology used and ranges from 0.25 to 5.0 million. The number of methods has also been used to determine the molecular weight (Burchard 1994; Ross-Murphy et al. 1998; Barth and Smith 1981; Vijayendran and Bone 1984). The molecular weight of galactomannan products was determined between 80,000 and 360,000 (Nuernberg and Bleimuellar 1981) and 2.2×10^6 (Vijayendran and Bone 1984). The molecular weight of clusterbean galactomannan was determined as 1.7×10^6 using light scattering method (Deb and Mukherjee 1963) and the viscosity of commercial gum preparations can vary depending upon its molecular weight.

The galactomannan structure of clusterbean has a certain degree of freedom of rotation, joining with two adjacent mannose units of polymer backbone. When the freedom of such rotation is restricted due to intra-chain hydrogen bonding of two adjacent mannose units and those of the galactose graft and mannose of the backbone, the polymer tends to have a straight, ribbon-like confirmation. The galactose and mannose sugar units have *cis* pair of hydroxyl groups at C-2, C-3 and C-3, C-4 positions, respectively, and due to the presence of these groups, galactomannans form strong hydrogen bonds with other polysaccharides.

The proportion of mannose to galactose units has been reported as 2:1 (Garti and Leser 2001) and several studies reveal the ratios in the range of 1.6:1–1.8:1 (Vijayendran and Bone 1984; Mathur and Mathur 2005). The X-ray diffraction studies by Marchessaut et al. (1979) revealed mannose:galactose (M:G) ratio of galactomannan to be 2:1 to 7:1. The greater branching of clusterbean gum is believed to be responsible for its easier hydration properties and superior hydrogen bonding (Arora 1989). It is also reported that these combinations are prominent in clusterbean systems and may have important role in viscoelastic behaviour of the solution (Gittings et al. 2001).

3.3 Properties

Clusterbean gum is white to yellowish white in colour, odourless and is available in different viscosities and different granulometrics (Rodge et al. 2012). Its natural properties depend on its behaviour in an aqueous medium. It dissolves in polar solvent on dispersion and forms strong hydrogen bonds, while in nonpolar solvents it forms weak bonds. The presence of salt plays peculiar role in the gelling mechanism and in the presence of salt the viscosity and gelling property are adversely affected. In case of a solution of low concentration, the salt addition facilitates the formation of inter-molecular aggregates due to alteration of charge density and conformation of the gum. Some reports suggest that the possible mechanism for this phenomenon is the disruption of intra- and inter-molecular associations causing expansion of chain conformation (Srichamroen 2007).

The chemical properties of galactomannan mainly depend upon their chemical features like chain length, abundance of *cis*-OH group, steric hindrance, degree of polymerization and additional substitutions (Pasha and Swamy 2008). The side groups that hold the main mannan chain and render ineffective non-covalent interactions are responsible for hydration of clusterbean gum in hot and cold water. The viscosity of the gum is affected by different factors like salt concentration, pH and type of agitation and is of temperature dependent. The viscosity of clusterbean gum ranged between 1.47 and 1.50 cp at pH 2–10 which is more or less constant revealing its non-ionic nature (Relan et al. 1991). The non-ionic nature of the molecule is responsible for the almost constant viscosity of solutions with pH from 1 to 10.5. At pH above 11, hydration is depressed and yields low viscosity. The optimum rate of hydration occurs between pH 7.5 and 9 and 1 %, and aqueous dispersion of good quality guar gum may show a high viscosity value of 10,000 cp (Parija et al. 2001). Rao et al. (1981) studied the effect of heat on the flow properties of aqueous clusterbean gum and found permanent loss in the apparent viscosity. Tiraferri et al. (2008) reported that the physical properties of guar gum do not change over the range of solution conditions although the temperature, ionic strength and pH of groundwater vary with location and fluctuate with time. The chemical and physical properties of clusterbean gum are discussed as follows:

1. The clusterbean gum hydrates rapidly in cold as well as hot water and gives highly viscous solutions. Hydration rate largely depends on particle size of the gum powder. Hence, for quick initial viscosity, very fine mesh guar gums are required (Glicksman 1969). The viscosity of clusterbean gum solution varies with shear rate. The viscosity of the gum increases linearly (Newtonian) with increase in concentration up to 0.5 %, and thereafter, the solution behaves as non-linear solutions due to complex surface attractions. The non-ionic nature of the molecule is responsible for almost constant viscosity of solutions in the range of 1–10.5 cp.
2. The specific colloidal nature of clusterbean gum gives the solution an excellent thickening agent and has good film-forming property.
3. It is stable over a wide range of pH and also functions at low temperatures.
4. The flowability and pumpability of any fluid can be improved with this gum as it has good water binding capacity.
5. It is a superior fraction reducing agent and provides resistance to oil, grease and solvent.
6. It retards ice crystal growth non-specifically by slowing mass transfer across solid–liquid interface.

The apparent viscosity parameters of clusterbean gum have been investigated by Naik (1980). It yielded very high viscosity at comparatively lower concentration and its compatibility with salts is exhibited over a wide range of electrolyte concentrations. In solutions form, the clusterbean gum shows decreased viscosity with increasing shear rate, and hence having a shear thinning behaviour and non-Newtonian flow in pseudoplastic sub-class (Zhang et al. 2005). The viscosity of the gum solution varies with shear rate and viscosity of 0.3 % gum solution changes slightly, whereas solutions of 1 % concentration or higher change markedly with change in shear rate. Viscosity depends on time, temperature, concentration, pH, ionic strength and type of agitation. Viscosity increases with time and short term heating, whereas it decreases with prolonged heating due to degradation effect on the structure of the gum (Srichamroen 2007). The viscosity of 1 % solution ranged from 3000 to 5500 cp, at 25 °C. The maximum viscosity of the gum dispersions is achieved at temperature of about 25–40 °C. The viscosity of a fully hydrated 1 % clusterbean gum solution varies almost directly with changes in temperature over the range of 20–80 °C. The viscosity increases linearly (Newtonian) with increase in concentration up to 0.5 %. Thereafter, the gum solution behaves as non-Newtonian (Joshi 2002). Whitecomb et al. (1980) reported that at sufficiently low and high shear rates, clusterbean gum solutions exhibit Newtonian behaviour while at intermediate shear rates, they yield pseudoplastic behaviour.

The gradual decrease in viscosity occurs after 24 h due to a drop in pH, fermentation and enzymatic hydrolysis. However, bacteriostatic and bacteriocidal preservatives in sufficient concentrations provide ample protection for gum solutions. Venkataiah and Mahadevan (1982) investigated the rheological properties of hydroxyl propyl and sodium carboxymethyl-substituted clusterbean gum in aqueous solution and found that the molecular distribution of clusterbean does not change by derivitization.

3.4 Estimation of Gum Content in Seed

A number of techniques were proposed for extraction, estimation and purification of clusterbean gum. The reliable and accurate technique involves the extraction and purification of galactomannan followed by alcohol precipitation and drying (Sikka and Johri 1969; McCleary and Matheson 1974). Besides this, colourimetric method (Mandal et al. 1989) is also used for gum estimation in clusterbean seeds. Patel (1958) isolated gum by loosening and removing husk of clusterbean seeds using chemical agent and also by cooking the seeds in water. McCleary and Matheson (1974) extracted gum by macerating the seeds as flour in 0.01 M $HgCl_2$ solution for 18 h followed by centrifugation and obtained a clear supernatant containing gum. A method for analysis of gum content in the seed has been described by high-performance liquid chromatography which serves a simple measure of gum contents in clusterbean and up to 20 samples/day can be processed (Hansen et al. 1992).

In another method, the round whole seeds are fed into a mill having two grinding surfaces travelling at different speeds. The seeds are split into two halves, i.e. endosperm with adhering hull and germ. The friable germ shatters and is shifted through, as fine material. The crude crack (endosperm + hull) is heated to soften the shell and is then fed into hammer mill, which can either shatter the hull away from the endosperm or into a mill or inactivates the toxic enzymes contained in seed germs (Murwan and Abdalla 2008).

Purification of the gum is generally done by repeated precipitation with alcohol. The isopropyl alcohol is mostly used in industrial processes of purification

(Meler 1965). Kendurkar (1970) refined clusterbean gum by adding 200 g gum to boiling solution of 0.2 % ammonium oxalate. Relan et al. (1991) purified the clusterbean gum using 80 % alcohol by repeated washings at low temperature in a soxhlet apparatus.

The available methods of estimation of gum are highly cumbersome and time consuming and thus become the limiting factor while screening large number of segregating populations/accessions. Pathak et al. (2010) developed a rapid screening method in clusterbean for identification of genotype having high gum content. Their study demonstrates that gum content regresses significantly on swelling weight of seed under different environments implying that by studying seed swelling, it may be possible to reject poor gum content lines.

3.5 Biochemistry of Clusterbean

The crude gum is greyish white powder and may contain small amount of proteinaceous matter. The major component of clusterbean gum is galactomannan (78–82 %), protein (4–5 %), crude fibre (1.5–2.0 %), ash (0.5–0.9 %), fat (0.5–0.75 %) and moisture (10–13 %) (Arora 1989). Glicksman (1969) also reported 78–82 % galactomannan, 10–13 % water, 4–5 % protein, 1.5–2 % crude fibre, 0.5–0.9 % ash and 0.5–0.75 % of fat in the commercial clusterbean gum. It consists of considerable fractions of Zn and Cu metals and protein to the tune of almost 5 %. It has abundant amino acids, viz., glycine, glutamic acid and aspartic acid. The clusterbean gum may contain small amount of proteinaceous matter which may arise from incomplete separation of the seed coat and endosperm prior to the milling. It also consists of small amount of various lipid soluble compounds including long-chain fatty acids, i.e. palmitate and linoleate. In addition, long-chain saturated acids and at least two hydroxy derivatives of fatty acids are also present (Gynther et al. 1982). The raw sample of clusterbean gum contains 0.07–1.42 µg fluoro acetic acid/gram, whereas pharmaceutical formulation contained 0.08 ppm fluoroacetate (Vertiainen and Pynther 1984).

Clusterbean seeds are good source of different natural compounds (Badr et al. 2013) including different flavonoids such as diadzein, genistein, quercetin and kempherol (Soehnlen et al. 2011). The seeds are low in methionine (0.22 mg/100 g), like most legume seeds (Murwan et al. 2012) and contains 2.47 mg/g total phenolic compounds and 2.85 mg/g tannins (Badr et al. 2014).

Arora et al. (1986) reported the presence of 5–6 % oil in clusterbean seed and suggested that the by-product of clusterbean could be utilized for extracting oil and making protein concentrate for human consumption. The fatty acid composition of clusterbean oil is comparable to that of sunflower oil (Joshi et al. 1981). The clusterbean oil contains 55.13 % linolenic acid against 51.8 % in sunflower oil. The total unsaturated fatty acids are 78.7 and 92 % in clusterbean and sunflower oil, respectively. The iodine value and refractive index of clusterbean oil are also well comparable with that of sunflower oil and can be used for edible purpose (Arora et al. 1986).

Kays et al. (2006) reported significant genotypic variability for total dietary fibre and soluble dietary fibre in the seeds of clusterbean. Significant variation for seed-derived daidzein, genistein, quercetin and kaempferol has also been reported (Wang and Morris 2007). Clusterbean seeds produced low amounts of daidzein and genistein and significantly more kaempferol as compared to soybeans (Wang and Morris 2007). Gallotannins, gallic acid, gallic acid derivatives, myricetin-7-glucoside-3-glycoside, kaempferol-7-glycoside-3-glycoside, kaempferol-3-rutinoside, kaempferol-3-glucoside, chlorogenic acid, caffeic acid and ellagic acid have been derived from clusterbean seeds. The polyphenolic content of clusterbean seeds depends on the stage of maturity and includes total phenol (1.26–0.69 %); gallic acid (0.49–0.72 %); gallotannins (0.5–0.21 %) and flavonols (0.13–0.23 %) on the basis of dry matter (Kaushal and Bhatia 1982). The amino acids (Khatta et al. 1988) and polyphenol content in the leaves also depend on the different growth stages and varieties of clusterbean.

Gum content in clusterbean has been found to vary from 19.1 to 34.1 % (Menon et al. 1970), 14.6 to 35.5 % (Dass et al. 1973) and 11.23 to 26.23 % (Tripathi and Srivastava 1975). The protein content in the seeds has been found to vary from 26.85 to 31 % (Misra et al. 1968), 21 to 33.25 % (Anonymous 1999) and 40.6 to 46.5 % (Mehrotra et al. 1975). Mehta and Rama Krishanan (1957) reported 4.2 % oil in seeds, whereas the maximum value of oil content (9.8 %) in the cotyledons has been reported by Shinde and Bhargava (1968). Joshi et al. (1990) also observed considerable variability in oil content (3.06–7.5 %) in seeds of various species of clusterbean. Ahmed et al. (2006) found 7.3 % moisture, 4.8 % ash, 9.3 % fibre, 2.3 % oil, 52.6 % protein and 23.7 % carbohydrate in clusterbean seeds. It was also noticed that the fatty acid composition of clusterbean oil is generally similar to the common edible oils (Singh and Misra 1981). Further, it was observed that *C. tetragonoloba* has higher oil content as compared to other *Cyamopsis* species (Arora et al. 1991). The clusterbean seeds contained 4.53 % ash, 3.32 % fat, 11.06 % fibre, 10 % moisture and 33.25 % protein. The most abundant minerals and fatty acids detected in seeds were iron and *cis*-linoleic acid (Badr et al. 2014). The methanolic extract of clusterbean seeds showed the presence of several metabolites, viz., 3-hydroxymyristic acid, octadecanoic acid and linolelaidic acid methyl ester (Badr et al. 2014).

3.6 Industrial Aspects of Clusterbean Gum

Clusterbean seed has attained an important place in industry because of its galactomannan-rich endosperm and protein. There are number of gum industries in India, which are playing a great role in domestic as well as in the international market and suppling 80 % of clusterbean gum in the world.

3.6.1 Processing

Clusterbean seed collected from farmers' field or stock market contains a lot of foreign particles like fine dust, mud, husk, stones, chaff, broken seeds, etc., and these materials are separated out before processing. The gum powder is extracted from the clusterbean seeds after a multistage industrial process. Various techniques are used for processing clusterbean depending upon the requirement of the end product and vary from plant to plant. In India, commercial production of gum is normally undertaken using mechanical process of roasting, differential attrition, sieving and polishing. Gunjal and Kadam (1991) suggested two processes, i.e., dry grinding and wet milling for splitting and dehulling of clusterbean.

3.6.2 Dry Milling Process

The clusterbean seed is mechanically split into two halves with the help of two grinding surfaces rotating at different speeds. The Jiwaji Industrial Research Laboratory, Gwalior obtained Indian patent in 1960 for this process. The split is heated to soften the hull over the endosperm and fed into the hammer mill to remove the hull by abrasion. The resultant endosperm is pulverized to remove residual hull and germ. A flow diagram for processing of clusterbean seed using dry milling process is given as follows:

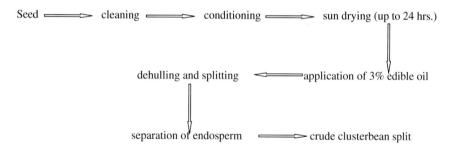

3.6.3 Wet Milling Process

It is one of the oldest methods of separation of clusterbean split developed by Patel (1958). The hull portion of the seed is loosened by cooking the seeds in water-containing $NaHCO_3$ and urea. The recovery in this process is 8–10 % higher as compared to dry processing method but the quality deteriorates (Gunjal and Kadam 1991). The flow diagram for processing of clusterbean seed using wet milling process is given as follows:

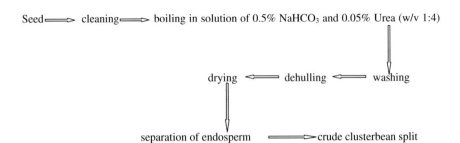

Seed ⟹ cleaning ⟹ boiling in solution of 0.5% NaHCO₃ and 0.05% Urea (w/v 1:4)

drying ⟸ dehulling ⟸ washing

separation of endosperm ⟹ crude clusterbean split

At present, both the dry and wet milling methods are not used by any industry for processing of clusterbean splits.

3.6.4 Conventional Method of Processing

There are more than 150 clusterbean split units in India (Sharma and Gummagolmath 2012) and about 125 units are located at the industrial area of Jodhpur (see Annexures). The seeds are graded in two or three sizes using cylindrical grader. The seeds are broken and the germ is separated from the endosperm by differential attrition. Two halves of the endosperm known as undehusked clusterbean splits are obtained from the seed. The split is then passes through a germ separator to remove the germ portion of the seed. Thereafter, the split is heated around 10–15 min through conduction to loosen the bond between hull and cotyledon to make dehulling easier. Usually, two or three dehulling process is required for complete removal of hull. After dehulling, the split is rapidly cooled and polished to get refined splits and the same is packaged. The clusterbean splits are usually of slightly concave or crescent shape. The refined splits are then treated and grounded into powders by a variety of methods and processing techniques. About 25 % of endosperm is converted into powder or broken during the dehulling process and get mixed with hull. The broken endosperm and powdered hull of clusterbean seed is the major by-product and is exploited as cattle feed.

A process of manufacturing food grade clusterbean gum is described here. The selected guar split is screened to clean and then soaked to prehydrate in a double-cone mixer. The soaked splits are passed through a flaker, dried and grounded to the desired particle size. Various grades of the powder are presented depending upon colour, mesh size, viscosity potential and rate of hydration (Chudzikowski 1971). Quality of food grade gum powder is defined from its particle size, rate of hydration and the microbial contains.

The clusterbean gum may contain husk residues represented by the acid-insoluble-matter criterion, proteins from the germ represented by the protein criteria, ethanol/isopropanol residues for washing or extraction solvent, microbiological

Table 3.1 The relationship between particle size and viscosity in clusterbean gum

Particle size (mesh)	Viscosity (cp)
80	Less than 3000
100	3000–6000
200	4000–9000
300	3000–5000

contamination, etc. This should be removed from the gum powder and should be further purified by dissolution in water, precipitation and recovery with ethanol or isopropanol. The ultrafine and uniform powder is used for different purposes because the particle size plays important role in the assessment of viscosity and produce solution of high viscosity (Table 3.1).

3.7 Guar Meal

Guar meal is a natural animal food supplement having high protein content (>47 %) and is also used as a binder for other feeds. During the gum production, clusterbean seed is split which yields high-protein germ portion and low-protein hull fraction as by-products. Hull of the seed available in powdered form is known as a *churi* and contains about 35 % protein, while the germ of the seed available in crystalline form is known as *korma* and contains about 50 % protein. These two fractions are then recombined/mixed to form guar meal which contains 35–47 % crude protein on dry matter basis (Ambegaokar et al. 1969). The clusterbean seed results in 62–68 % of guar meal depending upon the relative concentration of both the fractions. It is light, greyish tan colour material with beany odour and used for feeding to ruminants and livestock. The both forms of guar meal, i.e. *Churi* and *Korma,* have different proximate compositions (Table 3.2). The cotton seed and other such meal are often infected from *Salmonella*, Aflatoxin and *E. coli,* whereas guar meal is free from these infections. Guar meal is completely non-flammable and can be stored in silos, and does not create any problems in handling.

Guar meal is a good source of essential amino acids (Ramakrishnan 1957), but limiting in methionine, tryptophane and threonine (Ambegaokar et al. 1969). Some growth depression factors have also been reported in raw guar meal (Kawatra et al. 1969). The amino acid composition of clusterbean seed and its by-products has also been reported by different workers proving the quality of its protein content. Nagpal et al. (1971) reported that guar meals have sufficient amount of lysine and histidine. They found that tryptophan content was fair but methionine was deficient in guar meal and contained 3.4–5.7 % crude fat as ether extractives. A detailed account of amino acid estimated by different workers in different years is presented in Table 3.2.

Table 3.2 A detailed account of amino acid estimates from guar meal and its products

Amino acid	Free amino acid (g/16 g N)		Guar meal		Defatted guar meal (g/16 g N)	Guar protein (g/16 g N)	Guar protein concentrate (g/16 g N)
	Crude guar	48 h germinated seed	Coarse (% food)	Fine (% food)			
Alanine	0.12	1.2	–	–	3.08	3.65	3.42
Arginine	1.2	2.4	7.61	4.8	13.49	14.7	15.19
Aspartic acid	–	12.2	–	–	9.93	12.05	9.92
Cysteine	–	0.8	–	–	1.0	1.21	0.88
Glutamic acid	15.2	24.2	–	–	20.1	18.59	20.26
Glycine	0.07	1.23	–	–	7.66	7.15	7.36
Histidine	0.1	0.14	2.54	1.3	2.62	2.9	2.75
Isoleucine	2.1	10.8	–	–	2.9	3.12	3.26
Leucine	8.2	13.2	3.06	–	5.84	6.11	5.75
Lysine	9.1	12.3	3.05	2.43	3.95	4.14	4.3
Methionine	3.1	5.6	0.34	0.34	1.16	1.07	1.27
Phenylalanine	–	0.3	–	–	3.8	4.32	3.73
Serine	–	1.8	–	–	4.42	4.7	4.52
Threonine	–	1.9	–	–	2.67	3.84	2.83
Tryptophane	0.1	0.7	0.49	0.36	0.88	1.13	0.82
Tyrosine	0.5	1.2	–	–	3.73	3.94	2.77
Valine	5.3	7.5	–	–	3.07	2.83	3.29
Reference	Rama Krishanan (1957)		Nagpal et al. (1971)		Kaur et al. (1981)		

3.7.1 Toxic Constituents in Guar Meal

Clusterbean by-products may be economic for decreasing feed cost provided that the inclusion rates are kept lower. Guar meal contains certain anti-nutritional factors such as trypsin inhibitor (Hooper and Couch 1971), hemaglutinins (Arora and Joshi 1980), saponins (Curl et al. 1986), polyphenols (Kaushal and Bhatia 1982) and some unpleasant flavour components. These anti-nutritional factors have been reported to cause damage to liver, kidney and intestinal in mice and rats (Diwan et al. 2000). Gutierrez et al. (2007) suggested that guar meal and guar germ can be fed to high-production laying hens at levels up to 5 % of the diet without unfavourable effects on egg production, feed consumption, egg shell quality and solid egg components. Mohammad and Torki (2010) studied the effects of dietary inclusion of guar meal supplemented by β-mannanase on performance of laying hens, egg quality characteristics and diacritical counts of white blood cells and reported increased feed conversion ratio in the laying hens.

The higher concentration of guar meal in poultry diets cause diarrhoea, digesta viscosity, depresses growth rate, egg production and feed efficiency of laying hens

(Saxena and Pradhan 1974). Bakshi et al. (1964) found adverse effect of guar meal on chickens and hens. A number of adverse effects, viz., diarrhoea, depressed growth rate and increased mortality on feeding of clusterbean at relatively high levels, restrict its use in poultry feed (Patel and McGinnis 1985). However, numerous attempts have been made to detoxify the guar meal.

It was reported that guar dal treated with saline water and supplemented with trypsin do not depress the growth in rats. The trypsin inhibitor activity in toasted guar meal was decreased from 278 to 94 units/g by autoclaving the meal (Nagra et al. 1994), and it was further lowered through fermentation with *Aspergillus niger* and *Fusarium* like non-toxic fungi. Gupta and Vidyasagar Pradhan (1978) found two deleterious factors, viz., residual guar gum and trypsin inhibitor, in guar meal and suggested that toasting can improve its nutritive value. Rajput et al. (1998) suggested that autoclaving of raw guar meal had considerable destruction of haemagglutinins (84 %) and trypsin inhibitor (84 %) without change in saponin and phytate content. Toxicity study on partially hydrolyzed clusterbean gum has revealed that it is not mutagenic up to dose level of 2500 mg/day (Takahashi et al. 1994).

3.8   Derivatives of Clusterbean Gum

The derivatives of gum are prepared by modifications in the backbone of the galactomannan and impart desired properties, viz., increased solubility in water, ionic character, increased shelf life, solution clarity. These modifications open number of prospects where clusterbean derivatives can be used for rheological management of the system. Depolymerised guar gum, oxidized guar gum, hydroxy alkylated guar gum, borate cross-linked guar gum, carboxy methylated guar gum, acetates of guar gum, cationic derivatives of guar gum, sulphated guar gum, guar gum formate, guar gum acryl amide, reticulated guar gum, carboxy methyl hydroxy propyl guar gum, etc. are some of the commercially important derivatives of clusterbean gum (Swamy et al. 2006).

Clusterbean gum is depolymerized by oxidation either by peroxides in the presence of alkali or by reacting with strong acid between 40 and 70 °C temperature regime depending upon the end-use purposes. The depolymerized gum is mainly used in food and textile printing applications. Cross-linked guar gum is manufactured by the reaction of gum with sodium borate in alkaline medium and is mainly used in carpet printing and oil well drilling. The cationic derivatives of clusterbean gum are also known as quaternization of guar gum and are manufactured by reacting the gum with a quaternary ammonium compound to obtain guar-2-hydroxy-3-(trimethylammonio) propyl ether chloride or guar hydroxypropyl trimonium chloride. The cationic derivatives of clusterbean gum are the main choice of cosmetic and paper industries. Carboxymethylation of clusterbean gum is done by reacting the gum with sodium monochloro acetate in alkaline medium. These derivatives are very often used in textile printing and water-based paints. The reaction of clusterbean gum and propylene oxide in alkaline results in the hydroxypropylation of clusterbean gum and widely used in oil well drilling, paints, textile and cosmetics (Swamy et al. 2006).

3.9 Applications and Uses

Clusterbean gum is a natural vegetable gum and due to number of features, i.e. high natural viscosity due to its molecular structure, cold water solubility, rapidly hydration, freeze-thaw stability, etc., it has enormous applications in different industries. Clusterbean gum is used for upgrading the functions of starches due to their efficient thickening power. It reduces fractional pressure, losses of turbulent water flow and has good resistance to shear degradation. Highly refined guar gum is mainly produced for the food industry being used as a stiffener in soft ice cream, stabilizer for cheeses, instant puddings, whipped cream substitutes, meat binder, etc., while lower grade guar gum are used in cloth and paper manufacturing, oil well drilling muds, explosives, ore floatation and as host of other industrial applications. The large number of patents has been registered throughout the world on various uses of clusterbean. The patents for these products obtained by different countries are given at a website http://www.freepatentsonline.com/search.html. The patents have been grouped into seven categories at this website and are increasing daily. This site was accessed on 05.02.2015 and there were 108,608 patents related to clusterbean products in various countries or group of countries as given in Table 3.3.

The derivatization of clusterbean gum to introduce ionic and non-ionic sites has been used to extend the range of its application, solubility and gel forming properties (Swamy et al. 2006; Pasha and Swamy 2008). Its derivatives are used as crucial ingredients in about 100 products over diversified applications in different sectors of food, petroleum/gas exploration, explosives, pharmaceuticals, cosmetics, paper industry, textile, paints/distempers, aerial fire fighting, etc. Malkki et al. (1993) studied the effect of clusterbean gum on perception of sweetness and flavour. The clusterbean gum is a versatile product and enjoys wide usage in different industries, and there is still scope to explore its more usage. An account of important application of guar gum is described as under.

Table 3.3 List of patents in all fields of clusterbean

SN	Country	No. of patents
1.	US patents	25,290
2.	US patent applications	34,423
3.	EP documents	17,688
4.	Abstracts of Japan	301
5.	WIPO (PCT)	29,780
6.	German patents (beta)	1063
7.	Non-patent literature	63
Total		108,608

3.9.1 Paper Industry

Clusterbean gum provides better properties compared to other substitutes and gives denser surface to the paper used for printing. The major use of the gum and its cationic derivatives is the wet-end additive in pulp stages in newsprint, paper and board manufacture. The pulping process, which is designed to remove lignin and thereby produce a fibrous pulp, also removes a large part of the hemicelluloses normally present in wood which could contribute greatly to the hydration property of the pulp and the strength of the paper formed from the pulp. But during the past several years, the use of clusterbean gum in paper making has been reduced against cationic cellulose. For manufacturing a good quality paper and paper board, clusterbean gum and its modified gum are used to increase fine retention, impart dry strength, enhance surface and sizing (Anderson et al. 1986). The gum imparts improved erasive writing properties and increased hardness with better bonding strength.

3.9.2 Oil and Gas Well Drilling and Fracturing

The most significant use of clusterbean gum and its derivatives has been in petroleum production, especially the area of hydraulic fracturing, oil well fracturing, oil well stimulation and mud drilling. Clusterbean gum is used to thicken the fracturing fluid so that it can carry graded sand into the fractured rock. The carried sand serves as a proppant to keep the fracture open and create a route for oil or gas to flow to the bore well. In the oil field industry, clusterbean gum is used as a surfactant, synthetic polymer and deformer ideally suited for all rheological requirements of water-based and brine-based drilling fluids. High-viscosity gum products are used as drilling aids in oil well drilling, geological drilling and water drilling. Hydroxypropyl derivatives of guar gum are thermally more stable and are a choice for oil and gas well drilling and fracturing (Wu 1985).

3.9.3 Explosives

Water proofing, thickening and foam stabilizing properties are required in the preparation of explosives. Clusterbean gum is used as gelling agent for gel sausage-type explosives and pumpable slurry explosives (Ross and Eberhard 1987). Slurry explosives are generally based on various concentrated suspensions or solutions of nitrate salts, and therefore, gums to be employed must be compatible with high level of salts. Oxidized guar gum is the most commonly used polysaccharides to thicken explosives slurry. For making water-resistant slurry and gel explosives, the gum is mixed with ammonium nitrate and nitro-glycerine as a binding agent. Due to better swelling, water blocking and gelling quality, the gum maintains the explosive property in wet conditions also.

3.9.4 Mining Industry

Mining is one of the major industries which use clusterbean gum for extraction of metallic ores of potash, gold, copper and platinum group metals by froth floatation process (Pala et al. 2011). Some derivatives of clusterbean gum act to minimize water loss occurring in broken geological formations. It acts as a depressant in base metal floatation to block the absorption of other reagents onto the surfaces for talc or other insoluble gangue mined along with the valuable minerals.

3.9.5 Cosmetics

The unique cosmetic properties of clusterbean gum includes cold solubility, viscosity enhancing, solvent resistance film forming, protective colloid, wide pH range resistance, stability, non-toxic nature, safe, etc. So it is a choice of thickener, suspending agent, binder and emulsifier agent in various hair/skin care cosmetic products like creams, shampoos, premium quality soaps, lotions, conditioners and moisturizer. In the manufacturing of tooth paste, clusterbean gum binds the aqueous phase of the paste and is used in sizeable scale to impart flowing nature so that the paste can be extruded from the collapsible tubes with the application of a little force. In saving cream preparation, it does the same work besides providing stabilizing the system, imports slip during shaving and improves skin after shave (Chudzikowski 1971). In emulsion systems like cream and lotions, clusterbean gum prevents phase separation, sudden release of moisture, increase emulsion stability, prevent water loss and is used as protective colloid. It stabilizes the emulsion during freeze-thaw cycle, where the water phase condenses out of the system. In lotion, it provides additional spread ability and an agreeable feel (Srichamroen 2007). Cationic guar gum is used to thicken various cosmetics and toiletries products, especially to impart thickening, conditioning, foam stability, softening and lubricity. In aerosol dispensing aqueous liquid preparation as spray or mist, it reduces fog migration. The inert and compatible nature of clusterbean gum with the detergents makes it suitable for use in shampoo and cleansing preparation. Hair colourants contain clusterbean gum as thickener. It is also available in self-emulsifying grades and can be used to prepare dry face mask preparation (APEDA 1999).

3.9.6 Synthesis of Complexes

Clusterbean, being a high-molecular-weight polysaccharide, is a suitable surface modifier and its amalgamation with different nanoparticles has emerged as a new series of its application (Tiraferri et al. 2008). Clusterbean-based amalgamation of gas sensor gives response within 10 s and can be used for homeland security, gases leak detection in research lab and heavy metals removal from waste water (Gupta and Verma 2014). An optical sensor for ammonia detection was synthesized using clusterbean, silver nanocomposite (GG/AgNPs NC) and gold nanoparticles (GG/AuNPs NC) which will be able to detect low ammonia level in human (Pandey et al. 2012, 2013). Silver nanoparticles were synthesized using carboxymethyl guar grafted polyethylene oxide-co-propylene oxide which is an excellent polymer acting as reducing, stabilizing/capping agent and has an application in controlled drug release (Gupta et al. 2014). Clusterbean gum alkyl amine complexes with silver nanoparticles were synthesized for wound healing (Auddy et al. 2013). The new nanobiomaterial treated group gave faster healing and improved cosmetic appearance as compared to other silver alginate cream. The clusterbean gum is very effective in binding the silver clusters and restricts their size in the nanoregion (Biswal et al. 2009). Clusterbean is almost neutrally charged and renders its suitability for steric stabilization of the iron nanoparticles (Tiraferri et al. 2008). Clusterbean gum stabilized copper nanoparticles catalyst for cyclo-addition reaction has also been synthesized (Kumar et al. 2012). The clusterbean and nano-zinc oxide is used as an adsorbent for removal of Cr(VI) from aqueous solution and is capable to remove Cr(VI) from natural water resources (Khan et al. 2013). Clusterbean does not cause any risk to the environment and once introduced into the sub surface, it could enhance bioremediation.

3.9.7 Pharmaceutical Industry

Clusterbean gum or its derivatives are used in pharmaceutical industries as gelling, viscosifying, thickening, suspension, stabilization, emulsification, preservation, water retention/water phase control, binding, clouding, process aid, pour control for suspensions, antacid formulations, tablet binding, disintegration agent, controlled drug delivery systems, slimming aids, nutritional foods, etc. Its hydrophilic properties and the ability to form gel make it useful in gastric ulcer treatment. Studies revealed that a diet supplemented with clusterbean gum decreased the appetite, hunger and desire for eating (Butt et al. 2007).

In tablets, the gum is used as binder and increases the mechanical strength of tablets during pressing, and in jellies and ointments it is used as thickener. The depolymerized guar gum has been good bulking agent for dietetic food and as a food fibre, it is used in sugar and lipid metabolic control particularly in diabetic and heart patients (Saeed et al. 2012). Clusterbean gum powder is widely used in capsules as dietary fibre that decreases hypercholesterolemia, hyperglycemia and obesity (Dall'alba et al. 2013). The sufficient intake of clusterbean gum as dietary fibre helps in bowel regularity, reduction in total and LDL-cholesterol, control of diabetes, enhancement of mineral absorption and prevention of digestive problems like constipation and enhances bowel movement (Yoon et al. 2008). Some pharmaceutical companies are using clusterbean gum for making bandage paste and in dentistry formulations. The formulations like inhalation, injectable, beads, microparticles, nanoparticles, solid monolithic matrix films and implants also facilitate the use of clusterbean gum (Soumya et al. 2010). The high swelling characteristics of this gum sometimes hinder its use as a drug delivery carrier but it can be improved by derivatization, grafting and network formation and can be satisfactorily used for targeted drug delivery by forming coating matrix systems, hydrogels and nano-/microparticles (Prabhaharan 2011).

3.9.8   Textile Industry

On drying, solutions of clusterbean gum form flexible film which resists most organic solvents. However, these films are readily dissolved in water making them useful in textile sizing for protection of fibre during the weaving process. Clusterbean gum gives excellent film forming and thickening properties for textile sizing, finishing and printing. It reduces warp, breakage, dusting while sizing and gives better efficiency in production (Hebeish et al. 1986). Clusterbean gum derivatives like depolymerized, carboxymethyl and cross-linked guar gum are widely used in textile printing.

3.9.9   Food Industry

Clusterbean gum stands as one of the cheapest hydrocolloids in food industry and about 40 % of total clusterbean gum is presently used as food additive. The importance of this gum in food application is due to its various unique functional properties like water retention capacity, reduction in evaporation rate, alteration in freezing rate, modification in ice crystal formation, regulation of rheological properties and involvement in chemical transformation (Rodge et al. 2012). United States Food and Drug Administration (FDA) regulate the use of gums and classify these gums as either food additives or generally recognized as safe (GRAS) substances. The use of guar gum at a concentration not exceeding 2 % is allowed in food application. On addition of clusterbean gum, serum loss and flow values of tomato ketchup decreases which makes it a novel thickener for tomato ketchup (Gujral et al. 2002). Clusterbean gum has also been incorporated for therapeutic use in pizzas, biscuits and pastries (Griffith and Kennedy 1988).

Clusterbean gum prevents staleness and crumb formation in baked foods, provides unparallel moisture preservation to the dough, retards fat penetration, increases the dough volume, provides greater resiliency and improves texture and shelf life. In wheat bread dough, addition of clusterbean gum results in significant increase in loaf volume on baking (Cawley 1964). The depolymerized guar gum is used in the preparation of low calorie food. It enhances the creaming stability and control rheology of emulsion prepared by egg yolk (Ercelebi and Ibanoglu 2010). In beverages, it provides outstanding viscosity control, reduces calories value and improves the shelf life of beverages. In pastry fillings, it prevents weeping or release of moisture in the filling and keeps the pastry crust crisp (Miyazawa 2006). In dairy products, it thickens milk, yoghurt and liquid cheese products and helps to maintain the homogeneity and texture of ice cream and sherbets (Brennan and Tudorica 2008).

3.9.10 Animal Feed

Clusterbean provides a nutritious fodder and concentrate to the livestock and is used as a fresh or dry forage crop. The dry fodder of the crop is a rich source of protein but has low in total digestible nutrients (TDN), whereas the green forage contains about 16 % crude protein, 46 % TDN, 11–12 % digestible crude protein and 60 % dry matter digestibility on dry matter basis. These values vary with the stage of maturity. The clusterbean straw is quite nutritive and makes good fodder especially when mixed with wheat straw (Iqbal et al. 2015). Clusterbean forage was found more palatable as compared to the straw of *Cenchrus ciliaris* and *Vigna aconitifolia* and resulted in higher dry matter intake and live weight gain (Mathur et al. 2005). The clusterbean straw can be incorporated up to 70 % in the ratio of adult sheep to maintain them satisfactorily without any adverse effect (Singh et al. 2008) and the crop residues are also used for feeding camels in the arid regions of India (Bhakat et al. 2009).

3.9.11 Agriculture

Clusterbean gum works as water retaining agent, soil aggregation and anti-crusting agent for different agricultural applications and used as soil conditioner, soil additive, etc. and facilitates water holding capacity of soil. Among the number of organic polymers, clusterbean gum has anti-crusting action in the soil known to form a crust in the field in certain weather circumstances (Page and Quick 1979). The use of clusterbean gum has been reported in insecticide as a suspension agent and to improve their stickiness. To prevent the granule formation in fertilizers, it can be used as a coating agent along with fatty acid derivatives and kaolin. Clusterbean gum and starch form cross-linked products, when react with acrylamide and acrylic acid. One gram of this product has capacity to imbibe 800–1000 g water and form a thick gel. These products can be excellent source of water and can be put on the germinating seed and near the root caps of a growing plant during the water scarcity period for the survival of the plants (APEDA 1999). In Japan and Western countries, this product is being exploited effectively for such application. Clusterbean gum reduces the fraction of water through pipe during turbulent flow and helps in firefighting by increasing flow rate of water.

Clusterbean gum can be used in various ways, viz. flocculating agent in sodic soils, for improvement of infiltration rates in clayey soils, lining of water channels, seed coating with microbes, etc. Experiments have shown that the gum coating on the seeds enabled to hold maximum number of rhizobia or azotobacter cell and assist in better survival of inoculants (Mor et al. 1995). In an experiment of effect of different adhesives on nodulation and nitrogen fixation, the clusterbean gum increased number of nodules, plant mass and nitrogen fixation in pigeon pea. Seeds soaked in guar gum solution have shown better germination. Small amount of clusterbean gum powder put during the planting of plantlets in dry land farming helps in its growth because the gum can soak water from the soil and supplies it to roots slowly (APEDA 1999).

3.9.12   Medicinal Properties

Clusterbean gum reduces the absorption of sugars and lipid from the intestine. Thus, it helps in reducing blood glucose level in healthy and diabetic persons on oral consumption. Clusterbean gum normalizes the moisture content of the stool, absorb excess liquid in diarrhoea and soften the stool in constipation. It might help in decreasing the amount of cholesterol and glucose that is absorbed in the stomach and intestines. Clusterbean gum is a healthy soluble fibre which may help in lowering cholesterol level (Sultana et al. 2007) and decreases the risk of heart disease. When added to a meal, the gum provides a feeling of fullness and may help in reducing the body weight. The fibre in clusterbean slows down the digestion of a meal which lowers the glycemic index of the meal. The gum granules, tablet and capsule formulations are also available in different drug delivery forms and are used in numerous nutraceutical and pharmaceutical additives (Morris et al. 2004). Studies have shown that due to its water solubility and non-gelling behaviour, partially hydrolyzed gum decreased the symptoms in both constipation and diarrhoea predominant irritable bowel syndrome (Giannini et al. 2006) and is effective in treating number of diseases, viz., Crohn's disease, diabetes and colitis. Studies showed that clusterbean gum reduced the postprandial rise in blood glucose and insulin concentrations (Patel et al. 2014). The randomized controlled clinical trials showed that clusterbean gum supplementation decreased body weight and fasting plasma glucose and insulin concentrations in both healthy and metabolic syndrome patients (Cicero et al. 2010). Badr et al. (2014) investigated the anticancer, antimycoplasmal activities and chemical composition of clusterbean seeds and reported that the seeds extract possessed anticancer activities against PC-3, HCT116 and CACO-2 cell lines with half maximal inhibitory concentration.

3.9.13   Miscellaneous

Clusterbean gum is used in cement mortar formulation as water retention in construction industry. In water proofing and underground construction material, the gum is used as gelling agent (Jijima 1990). Clusterbean gum and its derivatives are used in metallurgy as binder (Ogbonlowo 1987). It is a choice of additive in number of preparations/application, viz., chalk for marking on stone, refracting insulating annuing composition, water-swellable clay compositions for binder and water absorber agents.

3.10 Export of Clusterbean Gum

The export of clusterbean from India was started in the form of seed but after the prohibition (www.indianembassy.org.), it confined to refined splits only. However, processed and semi-processed products of clusterbean seed, i.e. treated and pulverized clusterbean gum, refined clusterbean gum split and guar meal are allowed and being exported to about 100 countries. The major share of the clusterbean processed in India is exported as refined splits or powder. The world clusterbean market has matured and is increasing steadily every year. Export of clusterbean gum has increased by many folds in last 10 years. Most of the exports were destined to Western countries of USA, Germany, Japan, etc. The fluctuation in the export of clusterbean gum is always seen. During the oil embargo by the OPEC countries and price hike in year 1973 and 1979, export of clusterbean gum had increased due to its use in oil well drilling. The highest export was in the year 1981–1982 when the quantity of exported clusterbean gum was 1.14 lakh tones. Then a sharp decline was witnessed in its export. The export in year 1985–1986 was only 26,803 metric tonnes. This was the period when more than 80 % of the gum was used in oil drilling. In the recent years, export of modified/derivatives gums has also started. Later on, there was constant growth due to its increased use by the textile, paper industry and food processing industries. In the year 1990–1991, it was about 0.5 lakh MT which reached to 4.41 lakh MT in year 2010–2011 and 6.02 lakh MT in year 2013–2014. Similarly, clusterbean gum has emerged as the biggest foreign exchange earner. The export amount of Rs. 58 crores in year 1990–1991 had reached to Rs. 11,734 crores in year 2013–2014 which shows tremendous interest of the market. Under the export of farm sector, clusterbean gum is having the largest share. It is expected that export of the gum may increase significantly in the future. The export of clusterbean products from India to major importing countries for the last twenty seven years (1987–1988 to 2013–2014) is given in Table 3.4. USA, Germany and China are the major countries having maximum share of the Indian export of clusterbean products. These three countries imported 289.18 thousand MT (70.78 %) of clusterbean products from India and contributed Rs. 19,025 crore (89.38 %) to the export value during 2012–2013, whereas the USA alone contributed 60 % in terms of quantity and more than 81 % to the export value of clusterbean products exported from India during 2012–2013 and a remarkable increase in import of its products to USA has been noticed during the last three years.

Based on production and export, the apparent consumption of clusterbean gum can be found out by the trend in growth rated of 7–10 %/annum. Food industry

Table 3.4 Year-wise export of clusterbean gum from India

Year	Quantity (MT)	Value (Rs. in lakhs)
1987–1988	52,920.64	9899.55
1988–1989	35,422.12	7387.31
1989–1990	Not available	Not available
1990–1991	49,852.12	5814.53
1991–1992	61,426.86	9255.06
1992–1993	62,373.97	10,318.61
1993–1994	74,492.59	14,082.09
1994–1995	67,748.12	14,283.27
1995–1996	83,283.40	22,720.24
1996–1997	95,169.93	35,612.60
1997–1998	102,728.62	54,498.81
1998–1999	91,989.99	72,769.90
1999–2000	113,746.82	81,476.62
2000–2001	129,530.85	60,295.16
2001–2002	117,882.96	40,309.03
2002–2003	111,825.81	48,615.06
2003–2004	120,561.25	50,789.55
2004–2005	131,299.99	68,947.71
2005–2006	186,718.39	104,923.30
2006–2007	189,304.37	112,579.21
2007–2008	211,166.56	112,574.54
2008–2009	258,567.55	133,898.52
2009–2010	218,479.73	113,330.58
2010–2011	441,607.68	293,869.89
2011–2012	707,326.42	1,652,386.70
2012–2013	406,311.80	2,128,701.07
2013–2014	601,945.40	1,173,452.54

Source Annual export report of director general of commercial intelligence and statistics (DGCIS)

shares 30–40 %, and petroleum industry and mining industry share around 20–25 % of the total consumption of the gum. The shares of pharmaceutical, textile and paper industries range from 10 to 18 % of the total consumption of clusterbean gum. The industries are growing with a moderate rate indicating the good market potential. Latest statistics indicate that the clusterbean gum continues to enjoy an impressive growth in regard to annual production and increased exports. Figures from Director General of Commercial Intelligence and Statistics (DGCIS) show that clusterbean gum which has been regarded as the India's second largest agricultural export has overtaken rice, the main export crops and its export account for 18 % of India's total farm exports during 2010–2011. In the period 2012–2013, clusterbean gum export stood at $4.9 billion, followed by basmati rice ($2.6 billion) and cotton ($2.6 billion).

References

Ahmed MB, Hamed Rashed A, Ali Mohamed E et al (2006) Proximate composition, antinutritional factors and protein fraction of guar gum seeds as influenced by processing treatments. Pak J Nutr 5(5):481–484

Ambegaokar SD, Kameth JK, Shinde UP (1969) Nutritional studies in protein guar (*Cyamopsis tetragonoloba*). Indian J Nutr Diet 6:323

Anderson KR, Larson B, Thoresson HO et al (1986) (EKAAB POT Int. Appied WO 8500, 100 (CL D2H3I20) SE Appl. 84/306207, 35 pp (Cf. Guar Res Ann 5:44)

Anonymous (1999) Annual progress report. NBPGR, Pusa Campus, New Delhi

Anonymous (2015) Guar outlook 2015. CCS National Institute of Agricultural Marketing, p 64

APEDA (1999) A study on guar gum, p 121

Arora SK (1989) Guar endosperm-Its chemistry and utilization. In: Jasol FS (ed) Indian agro exports, pp 277–302

Arora SK, Joshi UN (1980) A thermostable haemolytic factor in guar (*Cyamopsis tetragonoloba*). Guar Newslett 1:13–14

Arora SK, Joshi UN, Jain V (1986) Lipids classes and fatty acid composition of two promising varieties of *Cyamopsis tetragonoloba* (L.) Taub. Guar Res Ann 4:9–10

Arora SK, Joshi UN, Arora RN (1991) Fatty acid spectrum of guar species. Guar Res Ann 7:57–59

Auddy RG, Abdullah MF, Das S et al (2013) New guar biopolymer silver nanocomposites for wound healing applications. Hindawi Publishing Corporation, Bio Med Res Int 1–9

Badr SEA, Sakr DM, Mahfouz SA et al (2013) Licorice (*Glycyrrhiza glabra* L.): Chemical composition and biological impacts. Res J Pharm Biol Chem Sci 4:606–621

Badr SEA, Abdelfattah MS, El-Sayed SH et al (2014) Evaluation of anticancer, antimycoplasmal activities and chemical composition of guar (*Cyamopsis tetragonoloba*) seeds extract. Res J Pharm Biol Chem Sci 5(3):413–423

Bakshi YK, Greger CR, Couch JR (1964) Studies on guar meal. Poult Sci 43:1302

Barth HG, Smith DA (1981) High-performance size-exclusion chromatography of guar gum. J Chromatogr 206:410–415

Bhakat C, Saini N, Pathak KML (2009) Comparative study on camel management systems for economic sustainability. J Camel Pract Res 16(1):77–81

Biswal J, Ramnani SP, Shirolikar S et al (2009) Synthesis of guar-gum-stabilized nanosized silver clusters with c radiation. J Appl Polym Sci 114:2348–2355

Brennan CS, Tudorica CM (2008) Carbohydrate based fat replacers in the modification of the rheological, textural and sensory quality of yoghurt: comparative study of the utilization of barley beta-glucan, guar gum and inulin. Int J Food Sci Technol 43:824–833

Burchard W (1994) Light scattering. In: Murphy R (ed) Physical techniques for the study of food biopolymers S B. Blackie Academic and Professional, London, pp 151–213

Butt MS, Shahzadi N, Sharif MK, Nasir M (2007) Guar gum: a miracle therapy for hypercholesterolemia, hyperglycemia and obesity. Crit Rev Food Sci Nutr 47:389–396

Casas JA, Mohedano AF, Garcia-Ochoa FJ (2000) Viscosity of guar gum and xanthan/guar gum mixture solutions. J Sci Food Agric 80:1722–1727

Cawley RW (1964) The role of wheat flour pentosans in baking. II. Effect of added flour pentosans and other gums on gluten starch loaves. J Sci Food Agr 15:834–838

Chakravorty JN, Chakravorty S, Bose SK (1979) On the molecular model and statistical dimensions of guar gum from theory of light scattering. Indian J Theor Phys 27:137–144

Chudzikowski RJ (1971) Guar gum and its applications. J Soc Cosmet Chem 22:43–60

Cicero AFG, Derosa G, Bove M et al (2010) Psyllium improves dyslipidaemia, hyperglycaemia and hypertension, while guar gum reduces body weight more rapidly in patients affected by metabolic syndrome following an AHA step 2 diet. Mediterr J Nutr Metab 3:47–57

Curl CL, Price KR, Fenwick GR (1986) Isolation and structural elucidation of a triterpenoid saponin from guar (*Cyamopsis tetragonoloba*). Phytochem 25:2675–2676

Dall'alba V, Silva FM, Antonio JP et al (2013) Improvement of the metabolic syndrome profile by soluble fibre—guar gum—in patients with type 2 diabetes a randomised clinical trial. Br J Nutr 110:1601–1610

Dass S, Arora ND, Singh VP (1973) Heritability estimates and genetic advance for gum and protein content along with seed yield and its component in clusterbean (*Cyamopsis tetragonoloba* (L.) Taub.). Haryana Agric University J Res 3:14–19

Dass PJH, Schols HA, De Jongh HHJ (2000) On the galactosyl distribution of commercial galactomannans Carbohyd Res 329(3):609–619

Deb SK, Mukerjee SN (1963) Molecular weight and dimensions of guar gum from light scattering in solution. Indian J Chem 1:413–414

Diwan FH, Abdel-Hassan IA, Mohammed ST (2000) Effect of saponin on mortality and histopathological changes in mice. East Mediterr Health J 6:345–351

Dodi G, Hritcu D, Popa MI (2011) Carboxymethylation of Guar Gum: Synthesis and Characterization. Cellulose Chem Technol 45(3–4):171–176

Doyle JP, Giannouli P, Martin EJ et al (2006) Effect of sugars, gallactose content and chain length on freeze thaw gelation of gallactomannans. Carbohydr Polym 64:391–401

Ercelebi EA, Ibanoglu E (2010) Stability and rheological properties of egg yolk granule stabilized emulsions with pectin and guar gum. Int J Food Prop 13:618–630

Garti N, Leser ME (2001) Emulsification properties of hydrocolloids. Polym Adv Technol 12:123–135

Giannini EG, Mansic DP, Savarino V (2006) Role of partially hydrolyzed guar gum in the irritable bowel syndrome. Nutr 22(3):334–342

Gittings MR, Cipelletti L, Trappe V et al (2001) The effect of solvent and ions on the structure and rheological properties of guar solutions. J Physical Chem 105:9310–9315

Glicksman M (ed) (1969) Gum technology in the food industry. Academic press, New York

Grasdalen H, Painter TJ (1980) NMR studies of composition and sequence in legume seed galactomannans. Carbohydr Res 81:59–66

Griffith AJ, Kennedy JF (1988) Biotechnology of polysaccharide. In: Chemistry Carbohydrate (ed) Kennedy JI. Publication, Oxford Science, pp 597–635

Guarda A, Rosell CM, Benedito C et al (2004) Different hydrocolloids as bread improvers and antistaling agents. Food Hydrocoll 18(2):241–247

Gujral HS, Sharma A, Singh N (2002) Effects of hydrocolloids, storage temperature and duration on the consistency of tomato ketchup. Int J Food Prop 5:179–191

Gunjal BB, Kadam SS (1991) CRC hand book of world food legumes, vol 1, pp 289–299

Gupta PC, Vidyasagar Pradhan K (1978) Guar as feed and fodder crop. In: Paroda RS, Arora SK (eds) Guar its improvement and management. The Indian Society of Forage Research, Hisar, India, pp 109–122

Gupta AP, Verma DK (2014) Guar gum and their derivatives: a research profile. Int J Adv Res 2(1):680–690

Gupta NR, Prasad BLV, Gopinath CS et al (2014) A Nanocomposite of silver and thermo-associating polymer by green route: a potential soft-hard material for controlled drug release. RSC Adv 4:10261–10268

Gutierrez O, Zhang C, Cartwright AL et al (2007) Use of guar by-products in high production laying hen diets. Poult Sci 86:1115–1120

Gynther J, Huhtikangas A, Kari I et al (1982) Free and esterified fatty acids of guar gum. Planta Med 46(9):60–63

Hansen RW, Byrnes SM, Johnson AD (1992) Determination of galactomannan (gum) in guar by high performance liquid chromatography. J Sci Food Agric 59(3):419–421

Hebeish A, El-Thalouth I, Ibrahim MA et al (1986) Technical feasibility of some thickeners in printing cotton reactive dyes. Ant Dyest Rep 75(2):22–29 (Cf. Guar Res Ann 5:46–47)

Hooper FG, Couch JR (1971) Trypsin inhibitor in guar meal. Fed Proc 30:641

Hui PA, Neukom H (1964) Properties of galactomannans. Tappi 47:39–42

Indianembassy (2008) http://www.indianembassy.org.cn/commercialwing/dynamiccontent.aspx?menuid=3&submenuid=0

Iorri W, Savanberg U (1995) An overview of the use of fermented food for child feeding in Tanzania. Eco Food Nutr 34:65–81

Iqbal MA, Iqbal A, Akbar N et al (2015) A study on feed stuffs role in enhancing the productivity of milch animals in Pakistan-existing scenario and future prospect. Global Veterinaria 14(1):23–33

Jijima S (1990) Temporary sealing composition for water proofing underground construction. Jpn Kokai Tokkyo Koho Jp 0253, 890 (9053, 8901), 5 pp. (Cf. Guar Res Ann 7:71)

Joshi UN (2002) Chemistry and nutritive value of guar. In: Singh NB (ed) Kumar D. Guar in India, Scientific Publishers (India), Jodhpur pp, pp 170–199

Joshi UN, Kaur A, Arora SK (1981) Fatty acid composition of guar oil. Guar Newsletter 2:43–44

Joshi UN, Arora SK, Arora RN (1990) Differential chemical composition of guar species. Guar Res Ann 6:38–40

Kamran M, Pasha TN, Mahmud A et al (2002) Effect of commercial enzyme (Natu grain) supplementation on the nutritive value and inclusion rate of guar meal in Broiler rations. Int J Poult Sci 1(6):167–173

Kaur A, Arora SK, Joshi UN (1981) Nutritive value of guar meal, protein isolate and concentrate. Guar Newslett 2:41–43

Kaushal GP, Bhatia IS (1982) A study of polyphenols in the seeds and leaves of guar (*Cyamopsis tetragonoloba* L. Taub) feed toxicity studies. J Sci Food Agric 33:461–470

Kawatra BL, Garch JS, Wagla DS (1969) Nutritional evaluation by rat feeding of preparations from guar seed (*Cyamopsis tetragonoloba*) including supplementation with lysine and methionine. J Agric Food Chem 17:1142–1145

Kays SE, Morris JB, Kim Y (2006) Total and soluble dietary fiber variation in *Cyamopsis tetragonoloba* (L.) Taub. (guar) genotypes. J Food Qual 29:383–391

Kendurkar SG (1970) Refining guar gum. Indian Patent 114:180. (Cf. Chem Abstr 74: 1279514)

Khan TA, Nazir M, Ali I et al (2013) Removal of Chromium (VI) from aqueous solution using guar gum–nano zinc oxide biocomposite adsorbent. Arab J Chem 1878–5352

Khatta VK, Kumar N, Gupta PC (1988) Chemical composition and amino acid profile of four varieties of guar (*Cyamopsis tetragonoloba*) seed. Indian J Anim Nutr 5:325–326

Kumar A, Aerry S, Saxena A et al (2012) Copper nanoparticulates in guar-gum: a recyclable catalytic system for the Huisgen[3 + 2]-Cycloaddition of Azides and Alkynes without additives under ambient conditions. Electronic Supplementary Material for Green Chemistry 1–11

Malkki Y, Heinio RL, Autio K (1993) Influence of oat gum, guar gum and carboxymethyl cellulose on the perception of sweetness and flavor. Food Hydrocoll 6:525–532

Mandal S, Kidwai MA, Dabas BS et al (1989) Evaluation of guar (*Cyamopsis tetragonoloba* (L.) Taub.) seeds for their gum and total protein content. Guar Res Ann 5:18–33

Marchessaut RH, Bullon A, Deslandes Y et al (1979) Comparison of X-ray diffraction data on galactomannans (from guar, tara and locust bean gum). J Coll Interface Sci 71:375–382

Mathur V, Mathur NK (2005) Fenugreek and other lesser known legume galactomannan-polysaccharides: scope for developments. J Sci Ind Res 64:475–481

Mathur BK, Patel AK, Bohra HC et al (2005) Acceptability, palatability and utilization of cluster-bean and dewbean fodder v/s buffel grass in sheep. J Arid Legumes 2(1):39–42

McCleary BV (1979) Enzymic hydrolysis, fine structure and gelling interaction of legume seed D-galacto-D-mannans. Carbohydr Res 71:205–230

McCleary BV, Matheson NK (1974) Galactosidase activity and galactomannan and galactosyl-sucrose oligosaccharide depletion in germinating legume seeds. Phytochemistry 13:1743–1757

McCleary BV, Dea ICM, Clark AH (1984) The fine structure of carob and guar galactomannans. In: Phillips GO, Wedlock DJ, Williams PA (eds) Gums and Stabilizers for the Food Industry 2:33–44

McCleary BV, Clark AH, Dea ICM et al (1985) The fine structures of guar and carob galactomannans. Carbohydr Res 139:237–260

Mehrotra ON, Tripathi RD, Srivastava GP et al (1975) Food value of some guar varieties. Indian J Agric Res 9:37–42

Mehta DR, Ramakrishanan CV (1957) Studies on guar seed oil. J Am Soc Oil Chem 34:459–461

Meler H (1965) Fractionation by precipitation with Barium Hydroxide Methods. Carbo Chem 5:45–46

Menon U, Dubey MM, Bhargava PD (1970) Gum content variation in gaur (*Cyamopsis tetragonoloba* (L.) Taub.). Indian J Hered 2:55–58

Misra DK, Bhan S, Prasad R (1968) Guara- a multipurpose summer legume. Allahabad Farming 42:239–244

Miyazawa T (2006) Hydrocolloid structures, which allow more water interactions through hydrogen bonding. Carbohydr Res 341:870–877

Mohammad E, Torki M (2010) Effects of dietary inclusion of guar meal supplemented by β-mannanase on performance of laying hens, egg quality characteristics and diacritical counts of white blood cells. Am J Anim Vet Sci 5(4):237–243

Mor S, Dogra RC, Dudeja SS (1995) Guar gum: an alternate adhesive for *Azotobacter* inoculation in cereals. Ann Biol 11(1):129–133

Morris JB, Moore KM, Eitzen JB (2004) Nutraceuticals and potential sources of phytopharmaceuticals from guar and velvet bean genetic resources regenerated in Georgia, USA. Curr Topics Phytochem 6:125–130

Mosha AC, Savanberg U (1990) The acceptance and food intake of bulk reduced weaning. The liganga village study. Food Nutr Bull 12:69–74

Mudgil D, Barak S, Khatkar BS (2014) Guar gum: processing, properties and food applications-A review. J Food Sci Technol 51:409–418

Murwan KS, Abdalla AH (2008) Yield and yield components of forty five guar (*Cyamopsis tetragonoloba*) genotype grown in Sudan. Nile Basin Res J 11(4):48–54

Murwan KS, Abdelwahab AH, Sulafa NH (2012) Quality assessment of guar gum (endosperm) of guar (*Cyamopsis tetragonoloba*). ISCA J Biol Sci 1(1):67–70

Nagpal ML, Agarwal OP, Bhatia IS (1971) Chemical and biological examination of guar meal (*Cyamopsis tetragonoloba*). Indian J Anim Sci 41:283–293

Nagra SS, Sethi RP, Chawla JS et al (1994) Improvement in nutritional value of guar meal by fungal fermentation. Indian J Anim Nutr 11(1):7–11

Naik SC (1980) Apparent viscosity characteristics of guar gum sols. In: Marrucci G, Nicolais L (eds) Astarita G., RheologySpringer Science + Business Media, New York pp, pp 341–345

Nuernberg E, Bleimuellar G (1981) Molecular weight of pharmaceutically available galactomannans products. Pharm Acta Helv 56:148–150

Ogbonlowo DB (1987) Potential of Jaguar in blast furnace pellet production. Trans Inst Min Metall Sect C. 96: C 186-C190 (Cf. Guar Res Ann 6:58)

Oke OL (1964) Hydrocyanic acid content and nitrogen fixing capacity of guar. In: Whistler RL, Hymowitz T (eds) Guar: agronomy, production, industrial use and nutrition. Purdue University Press. West Lafayette, Indiana, USA pp 125

Page ER, Quick MJ (1979) A comparison of the effectiveness of organic polymers as soil anticrusting agents. J Sci Food Agric 30(2):112–118

Pala S, Ghoraia S, Dasha MK et al (2011) Flocculation properties of polyacrylamide grafted carboxymethyl guar gum (CMG-g-PAM) synthesised by conventional and microwave assisted method. J Hazardous Materials 192:1580–1588

Pandey S, Goswami GK, Nanda KK (2012) Green synthesis of biopolymer–silver nanoparticle nanocomposite: an optical sensor for ammonia detection. Int J Biol Macromol 51:583–589

Pandey S, Goswami GK, Nanda KK (2013) Green synthesis of polysaccharide/gold nanoparticle nanocomposite: an efficient ammonia sensor. Carbohydr Polym 94:229–234

Parija S, Misra M, Mohanty AK (2001) Studies of natural gum adhesive extracts: an overview. Polym Rev 41:175–197

Pasha M, Swamy NGN (2008) Derivatization of guar to sodium carboxy methyl hydroxyl propyl derivative, characterization and evaluation. Pak J Pharma Sci 21(1):40–44

Patel RR (1958) Gum. Indian patent 61:005 (cf. Chem abstr 52:21190c)

Patel MB, McGinnis J (1985) The effect of autoclaving and enzyme supplementation of guar meal on the performance of chicks and laying hens. Poult Sci 64:1148–1156

Patel JJ, Karve M, Patel NK (2014) Guar gum: a versatile material for pharmaceutical industries. Int J Pharma Pharma Sci 6(8):13–19

Pathak R, Singh M, Henry A (2010) Breeding of gum content-a simplified technique for large scale screening of guar genotypes. J Food Legumes 23(3 & 4):243–244

Prabhaharan M (2011) Prospective of guar gum and its derivatives as controlled drug delivery system. Int J Biol Macromol 49(2):117–124

Rajput LP, Ramamani S, Saleem MA et al (1998) Chemical and biological studies on processed guar (*Cyamopsis tetragonoloba*) meal. Indian J Poult Sci 33(1):15–25

Ramakrishnan CV (1957) Amino acid composition of crude and germinated guar seed flour protein (*Cyamopsis psoralioides*). Experientia 13:78

Rana V (2013) Strategic report of guar Seed/gum. Dada Ganpati guar products (P) Ltd. http://dgguarpowder.com/downloads/files/n5208a9def0503.docx. Accessed 29 January 2015

Rao MA, Waltir RH, Cooley HJ (1981) Effect of heat treatment on the flow properties of aqueous guar gum solutions. J Food Sci 46(3):896–899

Relan PS, Kumar S, Garg V et al (1991) Rheological studies of guar gum by viscometer method. Guar Res Ann 7:25–29

Robinson G, Ross-Murphy SB, Morris ER (1982) Viscosity molecular weight retention-ship intrinsic chain flexibility and dynamic solution properties of guar galactomannan. Carbohydr Res 107:17–32

Rodge AB, Sonkamble SM, Salve RV et al (2012) Effect of hydrocolloid (guar gum) incorporation on the quality characteristics of bread. J Food Process Technol 3(2):1–7

Ross CA, Eberhard T (1987) Wet loading explosive. US4, 693, 763, 6 pp. (Cf. Guar Res Ann 6:59)

Ross-Murphy SB, Wang Q, Ellis PR (1998) Structure and mechanical properties of polysaccharides. Macromol Symp 127:13–21

Saeed S, Mosa-Al-Reza H, Fatemeh AN et al (2012) Antihyperglycemic and antihyperlipidemic effects of guar gum on streptozotocin-induced diabetes in male rats. Phcog Mag 8:65–72

Saini ML, Arora RN, Paroda RS (1981) Morphology of three species of genus *Cyamopsis*. Guar Newsletter 2:7–11

Saxena UC, Pradhan K (1974) Effect of high protein level on the replacement value of guar meal in layer ratio. Indian J Anim Sci 44:190–193

Sharma P, Gummagolmath KC (2012) Reforming Guar Industry in India: issues and strategies. Agric Econ Res Rev 25(1):37–48

Shinde SK, Bhargava PP (1968) Fatty oil from the guar seed cotyledons. Indian Oil Soap J 33:195–197

Sikka KC, Johri RP (1969) Mannose containing polysaccharides: a method for the extraction and estimation of gum of guar (*Cyamopsis tetragonoloba* (L.) Taub). Res Ind 14:138–139

Singh SP, Misra BK (1981) Lipids of guar seed meal (*Cyamopsis tetragonoloba*). J Agric Food Chem 20(5):907–909

Singh N, Arya RS, Sharma T et al (2008) Effect of feeding of clusterbean (Cyamopsis tetragonoloba) straw based complete feed in loose and compressed form on rumen and haemato-biochemical parameters in Marwari sheep. Vet Pract 9(2):110–115

Soehnlen MK, Tran MA, Lysczek HR et al (2011) Identification of novel small molecule antimicrobials targeting *Mycoplasma bovis*. J Antimicrob Chemother 66:574–577

Soumya RS, Ghosh S, Abraham EI (2010) Preparation and characterization of guar gum nanoparticles. Int J Biol Macromol 46(2):267–269

Srichamroen A (2007) Influence of temperature and salt on the viscosity property of guar gum. Naresuan University J 15(2):55–62

Sultana K, Islam F, Shamim SM et al (2007) Influence of guar on serum cholesterol. Pak J Pharmacol 24(2):37–40

Swamy NGN, Dharmarajan TS, Paranjyothy KLK (2006) Derivatization of guar to various hydroxyl alkyl derivatives and their characterization. Indian Drugs 43(9):756–759

Takahashi H, Yang SI, Fujiki M et al (1994) Toxicity studies of partially hydrolyzed guar gum. Int J Toxicol 13:273–278

Tiraferri A, Chen KL, Sethi R et al (2008) Reduced aggregation and sedimentation of zero-valent iron nanoparticles in the presence of guar gum. J CollInterface Sci 324:71–79

Tripathi RM, Srivastava GP (1975) Note on the gum content of some new strains of guar (*Cyamopsis psoralioides*). Indian J Agric Res 9:153–154

Venkataiah S, ad Mahadevan EG (1982) Rheological properties of hydroxylpropyl and sodium carboxymethyl substituted guar gum in aqueous solution. J Appl Polym Sci 27(5):1553–1548

Vertiainen T, Pynther J (1984) Fluoroacetic acid in guar gum. Food Chem Pharm 22:307–308

Vijayendran BR, Bone T (1984) Absolute molecular weight and molecular weight distribution of guar by size exclusion chromatography and low angle laser light scattering. Carbohydr Polym 4:299–313

Wang ML, Morris JB (2007) Flavonoid content in seeds of guar germplasm using HPLC. Plant Genet Resour Chara Util 5:96–99

Whistler RL, Hymowitz T (1979) Guar: agronomy, production, industrial use, and nutrition. Purdue University Press, West Lafayette

Whistler RL, Eoff WH, Doty DM (1950) Enzymic hydrolysis of guaran. J Am Chem Soc 72:4938–4939

Whitecomb PJ, Gutowski J, Warren WJ (1980) Rheology of guar solutions. J Appl Polym Sci 25(12):2815–2827

Winter WT, Chien YY, Bouckris H (1984) Structural aspects of food galactomannans In: Phillips GO, Wedlock DJ, Williams PA (eds) Gums and stabilizers for the food industry 2:535–539

Wu SR (1985) Alkaline refined gum and its use in improved well treating compositions. Eur Pat Appl EP 163, 27116 pp. (Cf. Chem Abstr 104:152129)

Yoon SJ, Chu DC, Juneja LR (2008) Chemical and physical properties. Safety and application of partially hydrolyzed guar gum as dietary fiber. J Clin Biochem Nutr 42:1–7

Zhang IM, Zhou JF, Hui PS (2005) A comparative study on viscosity behaviour of water soluble chemically modified guar gum derivatives with different functional lateral groups. J Sci Food Agric 85:2638–2644

Chapter 4
Cultivation

Abstract Clusterbean is an important leguminous annual crop of arid and semi-arid areas cultivated mainly in the north and north-west parts of India and the east and south-eastern part of Pakistan. The production of clusterbean in India is mainly confined to arid zones of Rajasthan and parts of Gujarat, Haryana and Punjab. It is sensitive to salinity at the seedling stage and cannot stand in water logging or stagnated water situation. The crop does not require extra attention for its cultivation and with little input, good production can be achieved. Problems associated with the breeding of clusterbean, its efficient plant types, detailed agronomic practices and varieties developed in India have been discussed at length in this chapter.

4.1 Introduction

Clusterbean is an important leguminous annual crop of arid and semi-arid areas cultivated mainly in the north and north-west parts of India and the east and south-eastern part of Pakistan. The production of clusterbean in India is mainly confined to arid zones of Rajasthan and parts of Gujarat, Haryana and Punjab. It is a photosensitive crop and grows well in specific climate conditions, which ensure a soil temperature around 21–25 °C for proper germination (Hymowitz and Matlock 1963). Long photoperiod with humid air during its growth period and finally short photoperiod with cool dry air at flowering and pod formation are the better environmental conditions for its cultivation. It has an indeterminate growth habit, remains vegetative and continues to flower and set pods from about 4 to 6 weeks following the emergence until terminated by either plant senescence or low temperature (Stafford and Hymowitz 1980). Clusterbean can be grown on variety of soil types from loamy to sandy soils. More compact soils disturb its root system with surface feeding nature and reduce nitrogen fixing bacterial activity.

Clusterbean is a rainy season crop but some varieties have been found to grow during March to June and other varieties during July to November as rainy season crop under South Indian climate conditions. In north-west India the crop is mostly

© Springer Science+Business Media Singapore 2015
R. Pathak, *Clusterbean: Physiology, Genetics and Cultivation*,
DOI 10.1007/978-981-287-907-3_4

grown as rainfed crop during the monsoon season, whereas, it is grown for green fodder and vegetable purposes, under irrigated conditions during hot summer pre-monsoon months. In southern India vegetable clusterbean is grown under irrigation throughout the year.

4.2 Efficient Plant Type of Clusterbean

It grows upright, reaching a height of 2–3 m with a main single stem with either basal branching or fine branching along the stem. Clusterbean is a robust, bushy crop and has well-developed tap root system. Stems and branches are angular, grooved, forked hairs and sometimes greyish green. It has pointed saw-toothed, alternate, tri-foliate leaves with small purple and white flowers borne along the axis of spikelet. It bears hairy pods in clusters of 4–12 cm length, each pod with 7–8 seeds. Seed is hard, flinty, flattened, ovoid and about 5 mm long, white, grey or black in colour. Studies suggest that branched variety having more leaves and less lignin content in the stem had better digestibility and intake as compared to the branched variety (Singh et al. 1976). Yadav et al. (1989) also reported better performance of the branched genotypes over unbranched. A model clusterbean plant should possess few branches, high number of clusters with pods, bold seeds and long peduncle for higher seed yield (Chaudhary and Singh 1976), besides this early maturing, determinate growth habit, photo insensitivity, moderate number of branches and height are also expected as efficient plant type for higher yields of this crop.

4.3 Crop Cycle

Clusterbean takes about 90–110 days from sowing to harvesting. However, the crop cycles fluctuate from 60 to 90 days for determinant varieties and 120–150 days for indeterminate varieties (NRAA 2014). The sowing is done by the first to second week of July after rainfall or it may be extended up to August depending on the onset of monsoon. Germination takes place within 4–6 days after sowing. The germination is dependent on different factors, viz. seed viability, seed size, seed vigour, soil type, soil moisture, soil temperature and relative humidity. The bolder seeds give higher germination, better shoot length and higher dry matter production as compared to small or medium sized seeds (Renugadevi et al. 2009). Clusterbean is not able to germinate well under high saline and submergence conditions. Pre-soaking of the seeds in good quality water for 2 h followed by half an hour shade drying may increase germination percentage. In general, flowering on the crop starts after 40–60 days and pod formation takes place after 50–70 days from the date of sowing. Pod matures in 80–90 days. The harvesting of the crop is started in 90–110 days from the date of sowing when 90 % pods are matured depending on the variety, soil and climatic conditions.

4.4 Climate

Clusterbean is a photosensitive crop and requires specific climatic condition to grow. It tolerates high temperatures and dry conditions and is well adapted to arid and semi-arid climates. For proper germination the ideal soil temperature should be around 20–25 °C, long day period for vegetative growth and short day period for flowering and pod formation. At planting time, soil temperatures should be above 21 °C for rapid establishment. The optimum temperature for root development is 24–30 °C (Daisy 1979), but it can tolerate temperatures ranging from 10 to 45 °C. Seedling emergence occurs at a base temperature of 14.6 °C (Angus et al. 1981). The crop can tolerate high temperature even up to 40 °C but it should not be below 15 °C during the growing season (Tyagi et al. 1982). Towards the end of the season, the weather must be dry with abundant sunshine. The plant is tolerant to shade, but susceptible to frosts and about 110–130 frost-free days are required to attain the maturity (FAO-Ecocrop 2007). Although the optimum annual rainfall for clusterbean fluctuates between 400 and 800 mm (Yousif 1984), it can be grown without irrigation even in areas with 250 mm of annual rainfall (Undersander et al. 1991). In arid condition it is grown as rainfed crop and requires 300–400 mm rainfall in 3–4 spells. Pathak and Roy (2015) studied the climatic responses, environmental indices and interrelationships between qualitative and quantitative trains in clusterbean under arid conditions and found that the crop gave good yields with temperature regime of 36.7 °C, sunshine of 11 h and water stress during 32–35th meteorological weeks of cropping season under the rainfed and natural climatic conditions. These climatic conditions also favoured endosperm development and so the gum content in clusterbean seed.

Clusterbean can be grown on different types of soils but light-textured sandy soils are more suitable for the crop. It performs well on fertile, medium fertile, medium textured, sandy loam alluvial soils, unlogged and well-drained subsoil, but it cannot tolerate heavy black soils (FAO-Ecocrop 2007). Elsayed (1994) reported that soil salinity significantly decreased nodulation, pod formation and seed yield in clusterbean. Francois et al. (1990) found that soil salinity up to 8.8 dS m^{-1} did not affect the crop, but per unit increase above 8.8 dS m^{-1} reduced yield by 17 %, which placed it in the moderately tolerant crops category.

The researchers have also explored the possibility of cultivating the crop in non-traditional seasons. The average grain yield was almost three times more in summer compared to rainy season yield, besides this gum content was also higher in summer crop suggesting that the summer clusterbean would be of immense importance (Anonymous 2013). The crop cultivated during summer season showed favourable results over the crop under rainy season in terms of maturity period, plant height, number of pods/plant and number of seeds/pod. The cultivation of the crop in summer season may help in increasing crop yield, gum content and viscosity over the rainy season cultivated crop. It may be due to favourable climatic conditions, i.e. higher temperature, lower humidity, sunny sky and controlled availability of soil moisture, less diseases and insect pests infection

(Satyavathi et al. 2014). The cultivation of the crop in non-traditional season may provide new avenues, economic use of available irrigation water and increasing cropping intensity in arid and semi-arid regions.

4.5 Sowing Method

Two or three ploughings of the field is required for good seed germination, plant growth, soil aeration and root development. Each plough should be followed by planking. At the time of sowing the field should be well drained and weed free. Traditionally, the crop is sown by broadcasting method in India and no seed bed is prepared. In this method inter and intra row spacing is not followed which creates problem in various intercultural operations, viz. hoeing, weeding and removing excessive water from the field. The crop sown under proper line and row spacing may help in enhancing the productivity of the crop. The sowing using line method is normally done by seed drill which ensures sowing with proper spacing and depth. This method results in good germination, proper weed management and drainage of excess water in the field. Prajapati et al. (2004) optimized the plot size for rainfed clusterbean using uniformity trial data and suggested that a plot of 10.8 m^2 having shape of 4 m length (East–West) and 2.7 m cross width (6 rows in North–South) is optimum size and shape for rainfed clusterbean crop experiment.

4.6 Sowing Time

The time of sowing plays a crucial and important role in the growth and subsequently in the seed yield of the crop. Sharma et al. (1984) studied the effect of dates of sowing on yield and quality of clusterbean suggesting that maximum seed yield can be realized with the crop sown on 5 July as compared to 20 June, 20 July and 5 August. Tiwana and Tiwana (1992) suggested that higher mean seed yield can be obtained with crop sown on 30 June with 30 kg seed/ha at 30 or 45 cm spacing in Bhatinda, Punjab region under rainfed conditions. There are number of reports which indicate that crops sown on 10 July recorded maximum seed yield, protein and gum content (Dhukia and Singh 1988; Yadav et al. 1992; Taneja et al. 1995). Clusterbean grown in the end of June or early July with 20 kg seeds/ha and a row spacing of 30–45 cm gave the highest seed yield under irrigated conditions (Tiwana and Tiwana 1993). In general, in most of the clusterbean growing regions, the optimum time for the sowing of crop is first week of July to get higher seed yield (Table 4.1). The summer crop in northern India may be sown in the month of March, while it can be sown at any time between February and October in southern India.

Table 4.1 Optimum inter and intra row spacing for sowing of clusterbean seed

Sowing time	Plant type	Inter row spacing (cm)	Intra row spacing (cm)	Plant popula-tion (00,000/ha)	Seed yield (kg/ha)
Normal (1–15 July)	Branched	45	10–15	1.5–2.2	10–12
	Unbranched/ sparsely branched	30	10–15	2.8–3.3	12–15
Late (July 25–August 5)	All varieties	30	8–10	3.3	15–20

4.7 Spacing

The plant to plant and row to row distance ensure proper utilization of different natural inputs, viz. nutrition, moisture, sunlight and minimize the plant competition for growth. The spacing depends on the optimum plant stand and varies for different regions having varied rainfall intensities. The branched genotypes with the row spacing of 30 cm gave significantly higher seed yield due to more number of plants and higher number of pods/plant (Bhadoria and Kushwaha 1995). Optimum inter and intra row spacing for branched and unbranched genotype/ varieties of clusterbean is shown in Table 4.1. Row to row and plant to plant distance of 45 and 15 cm, respectively, gives optimum seed yield however, it may be reduced for late sown crops as well crop sown under poor soil fertility conditions (Bains and Dhillon 1977).

Yadav et al. (1992) suggested that the different row spacing had no major effect on seed yield, although, they found more number of pods and higher seed yield/ plant with 45 cm of row spacing. Rana et al. (1991) recorded higher grain yields with 30 cm inter row spacing under late sown conditions for different varieties and suggested that 15 cm intra row spacing is more suitable for branched and 10 cm for unbranched varieties. Reddy and Reddy (2011) studied the effect of planting geometry and fertility level on seed yield of clusterbean and reported that the planting pattern of 45 × 10 cm gave higher plant height and number of branches/ plant, however, it was comparable with 30 × 10 cm planting pattern and significantly superior to 30 × 5 cm planting geometry. They recommended the adoption of 30 × 10 cm spacing with 50 % recommended dosage of fertilizers (RDF) for scarce rainfall zone of Andhra Pradesh.

4.8 Seed Rate

Seed rate mainly depends upon the variety, purpose, time of sowing, availability of soil moisture and spacing of rows in which the crop is to be sown. Higher seed rate is required for broadcasting, late sown crop and crop sown under saline or alkaline soil. Generally, 10–12 kg seed/ha is recommended for spreading-type varieties whereas 15–16 kg/ha for unbranched varieties of clusterbean. The crop grown for grain production requires about 20 kg seed/ha, whereas for fodder and green manuring 40 kg seed/ha would be sufficient. Studies suggest that seed rates of 20 and 30 kg were at par and were superior to 10 kg seed rate/ha (Singh et al. 1978; Singh and Singh 1989).

4.9 Seed Treatment and Inoculation with Bacterial Culture

Since clusterbean is a leguminous crop and rhizobium bacteria is symbiotically associated with its root in the form of nodules, it converts atmospheric free nitrogen into the available form of fertilizer. For areas which are prone to disease outbreak, it is essential to inoculate the seeds before sowing for success of crop. Seed is treated with Ceresan or Thiram at the rate of 3 gm/kg seed to destroy the spores of dry root rot fungus resting on the seed coat. Seeds can be treated with Imidachlorpid at the rate of 6 ml/kg seed to control sucking pest like Jassids and Aphids. Seed immersed in hot water at 50 °C for 10 min followed by drying at room temperature before sowing helps to kill the fungal mycelium and inactivate their spore. For inoculation of the seed, 10 % sugar solution is prepared in boiling water and is allowed to cool. After cooling, 3–4 packets of bacterial culture are mixed properly with the solution to make a thin paste. This paste is coated over the seed. The seed is dried under shed for 30–40 min before sowing in the field. Kumhar et al. (2012) studied integrated nutrient management in clusterbean and reported that the application of 100 % recommended dosage of nitrogen through Urea + Rhizobium + PSB recorded significantly higher plant height, number of nodules/plant, green weight of nodule/plant, number of pods/plant, seed yield/plant, 100-seed weight and dry matter/plant at harvest.

4.10 Germination and Seedling Growth

Bolder seeds give higher germination, increased root and shoot length and higher dry matter as compared to medium and small seeds of clusterbean (Kalavathi and Ramamoorthy 1992). Renugadevi et al. (2009) reported a positive association between seed size and seed and seedling quality characters in clusterbean. The increase in seed size led to increased root and shoot length and higher dry matter production.

The seed of clusterbean takes six days for complete germination and the optimum temperature for its germination is 30 °C (Zade et al. 1990). Studies for obtaining prompt and uniform germination suggest that a constant temperature of 30 °C, seed treatment with dilute sulphuric acid (Musil 1946), scarification for 5 min (Hymowitz and Matlock 1964), pre-soaking of seed in NaCl (Singh et al. 1976) and sowing depth of 15–30 cm (Zheng et al. 1980) are the practical means for rapid, better and constant germination in clusterbean. The seed of clusterbean is not able to germinate at very low soil water potentials, high salinity and submergence conditions (Datta and Dayal 1988).

4.11 Clusterbean Under Crop Rotation System

Every crop do not require the nutrients in same proportion, so by growing crops in rotation the fertility of the soil can be utilized more evenly and can be prevented from nutrient depletion. Clusterbean has been identified as the most preferred crop under crop rotation. Clusterbean–wheat crop rotation is more popular in the eastern parts of Rajasthan and eastern–southern parts of Haryana and has maximum water use efficiency (Singh et al. 1998). Saxena et al. (1997) recorded higher grain yield of pearl millet when it was grown after clusterbean as compared to continuous cropping of pearl millet in a particular field. Inclusion of clusterbean in the pearl millet-based crop rotation for 1, 2 or 3 years had improved various parameters of biological productivity of soil over a continuous fallow or continuous pearl millet resulting in a significant increase in pearl millet production.

4.12 Clusterbean Under Intercropping System

The intercropping system improves soil fertility, facilitates yield advantage of the companion crop and provides insurance against crop failure due to any biotic or abiotic hindrance. Farmers are in practice to grow clusterbean in mixture with other legumes, cereal, oil seed crop, etc. in varying proportions. Intercropping of pearl millet and clusterbean is an advantageous practice during the good and subnormal years but it remained unprofitable in drought conditions (Daulay et al. 1998). Further, it was revealed that in intercropping with pearl millet–clusterbean, the higher returns were obtained with sole crop of clusterbean indicating that its intercropping with pearl millet has no economic significance (Yadav and Joon 1993). Clusterbean proved superior and gave more grain yield when two rows of clusterbean were sown in between paired rows of pearl millet (Bhati and Manohar 1989; Singh 1981). The study of different pearl millet-based cropping systems in drought conditions revealed that strip cropping comprising of clusterbean, gave higher yields and covered more risk as compared to other systems (Singh and Joshi 1994).

Clusterbean intercropping had beneficial effects on grain yield of maize (Singh and Kaushik 1997) and sorghum (Singh and Ahuja 1990). The rows of rainfed maize coupled with one row of clusterbean had higher grain yield and net profit as compared to sole maize crop (Chauhan and Dungarwal 1982). Intercropping of two rows of clusterbean in castor grown in paired rows showed significant increase in land equivalent ratio, yield and net returns in comparison to castor in pure stands (Singh and Singh 1988), on the contrary Venkateswarlu and Subramanian (1990) and Reddy and Venkateswarlu (1989) found clusterbean as highly sensitive in mixed stand with castor. The intercropping of clusterbean with oilseeds, viz. castor, sunflower, groundnut, sesame, pigeon pea, etc. is not always beneficial (Yadav et al. 1997a).

4.13 Fertilizer Management

Clusterbean is a leguminous crop and as such it does not require much nitrogen but a modest dose of nitrogen helps in the stimulation of plant growth in its early stage and improve nodulation, growth and nitrogen fixation (Atwal and Sidhu 1964). Balanced fertilization along with sound crop husbandry offers a great scope for increasing productivity of clusterbean (Sharma and Singh 2005). Number of studies suggest that higher grain/straw yield, water use efficiency, gum content and net return can be achieved with 20 kg nitrogen/ha as compared to no nitrogen (Yadav et al. 1991a, b). However, 30 kg N/ha has also been recommended for higher seed yields, nitrogen uptake and concentration of nitrogen in seed and straw (Singh et al. 1993). It has been found that 15–20 kg nitrogen/ha gives good start to the crop. Application of excessive amounts of nitrogen results in slow nitrogen fixation process.

Phosphate has been considered as the most useful fertilizer but its response may vary from one location to another (Taneja et al. 1981). Numerous reports indicate that the application of phosphorus improves the growth, root dry matter, leaf area index, nodulation, net assimilation rate, bolder seed size and yield of clusterbean (Singh and Singh 1989; Yadav et al. 1990). In some studies a dose of 40 kg P_2O_5/ha was found superior for different crop parameters including yield attributing traits, viz. plant growth, seed and stalk yield, dry matter accumulation, crop growth rate, leaf area index and yield contributing characters, viz. number of pods/clusters, number of pods/plant, number of seeds/pod, 1000-seed weight and seed yield/plant (Shivran et al. 1996). However, in some studies, it is suggested that phosphorus has no response beyond the dose of 20 kg P_2O_5/ha (Taneja et al. 1981) or 30 kg P_2O_5/ha (Gill and Singh 1981) under rainfed conditions.

The fertilizer level of 25:94:75 (NPK kg/ha) gave significantly higher plant height, number of pods/plant, pod weight/plant, number of seeds/pod and seed yield. Further, this fertilizer dose induced early flower initiation and lowered the day to 50 % flowering (Palankar and Malabasari 2014). Ramawtar et al. (2013) studied the effect of NP fertilizers, vermicompost and sulphur on growth, yield attributes, yield and quality of clusterbean and reported that the combined application of NP fertilizers at 75 % of RDF and vermicompost at 2 t/ha significantly improved the pods/plant, seeds/pod, pod length, gum content and seed yield of clusterbean. Fertilization of the crop by 70 kg phosphorus and 25 kg nitrogen produced higher percentage of normal seedlings as compared to unfertilized crop (Madalgeri and Rao 1989).

Table 4.2 Requirement of optimum level of micronutrients for better growth of clusterbean

Micronutrient	Optimum level (ppm)
Calcium	180–200
Magnesium	5–10
Sulphur	30–35
Boron	0.1–0.3
Copper	0.1–0.2

Shekhawat et al. (1996) found gypsum as better source of sulphur than pyrite and recorded higher number of seeds/pod and branches in clusterbean. Application of 25 and 50 kg sulphur/ha increased the pods/plant by 19.7 and 31.4 %, grains/plant by 8.9 and 11.6 % and test weight by 14.9 and 17.1 % over control, respectively (Sharma and Singh 2005). The significant improvement in yield obtained under sulphur fertilization seems to have resulted in the form of increased concentration of sulphur in various parts of clusterbean that helped to maintain the critical balance of other essential nutrients and resulted in increased metabolic processes in the plants. Yadav et al. (2012) studied the influence of phosphorus and sulphur on yield and micronutrient uptake by clusterbean and observed significant increase in the grain and straw yield with increase in level of P and S individually as well as in various combinations.

Direct application of 50 kg $ZnSO_4$/ha is found better for clusterbean yield under acute zinc deficiency soils (Takkar et al. 1973). Application of $ZnSO_4$ at 10 kg/ha and 20 kg/ha has also been recommended depending upon its deficiency. Foliar spray of 1 kg $ZnSO_4$/ha at 45 DoS has better effect than soil application of 5 or 10 kg $ZnSO_4$/ha. Nandwal et al. (1990) reported that three sprays of 0.5 % $ZnSO_4$ at 15 days interval is more effective and increases nodules, protein, carbohydrate content and number of pods, seed weight and leaf area. The study reveals that zinc alone or in combination with other micronutrient (Cu, Mn and Fe) significantly increases dry matter yield. The optimum level of various micronutrients is presented in Table 4.2.

Reddy et al. (2014) suggested that clusterbean should be supplied with the 75 % of RDF and remaining 25 % RDF through vermicompost along with biofertilizers (Rhizobium at 25 g/kg seed + PSB at 5 kg/ha) for getting optimum growth and higher pod yield. Chavan et al. (2015) observed the effects of organic and chemical fertilizers on clusterbean and reported significant increase in growth and pods yield in plants treated by vermicompost fertilizer followed by chemical fertilizers.

4.14 Irrigation Management

The crop is highly sensitive to water logging throughout development period. So, drainage of water is very important during rainy seasons. In general, clusterbean requires less number of irrigation. In rainy season, it is normally grown without any irrigation depending upon the rainfall distribution. While, summer season crop is irrigation dependent and 2–3 irrigation is enough to raise the crop. Studies suggest that one life-saving irrigation at 60 days of sowing increased the grain yield significantly over no irrigation (Meena et al. 1991). Other workers also recommended 1–2 irrigations for higher crop yield (Singh et al. 1998). The crop does not require irrigation at the early vegetative stage but irrigation at flowering and pod formation stage helps the crop in higher yields and productivity.

4.15   Weed Management

Clusterbean, being a rainy season crop, has to compete for moisture, nutrients and space with a large number of weeds come up during the rainy season and considerable reduction in crop yield has been recorded (Bhadoria et al. 1996). Reports indicate that weed control alone is responsible to increase the seed yield by 61–68 % in the absence of any input (Yadav et al. 1993). The most common weeds that infest clusterbean growing regions in India are *Cyperus rotundus*, *Trianthema portulacastrum*, *Digera arvensis*, *Tribulus terrestris*, *Amaranthus viridis*, *Cyanodon dactylon*, *Celosia argentina*, *Phylanthus niruri*, *Echinochloa colonum*, *Pulicaria wightiana*, *Portulaca oleracea*, *Digitarea sanguinalis,* etc. (Bhadoria et al. 1996). Studies show that critical period of crop–weed competition lies between 20 and 50 DoS (Yadav 1998), from 15 to 30 DoS on loamy sand and sandy loam (Yadav et al. 1997b). The mechanical methods of weed control are more suited in crop production because along with weed control it conserves the soil moisture of the field and are eco-friendly and are economic (Bhadoria et al. 1996; Yadav et al. 1997c). Early weed control is the basic necessity to get full-yield potential of the crop. Therefore, the first interculture operations should essentially be done within 25–30 DoS. Bhadoria and Jain (2005) studied the crop–weed competition in clusterbean under rainfed condition and reported that the weed-free plot had the maximum seed yield and weed control efficiency.

Yadav et al. (1998) found fluchloralin as efficient weed control in clusterbean which can be incorporated in the soil before sowing at the rate of 1.5 kg active ingredients/ha. But use of this herbicide at the pre-emergence stage does not work properly (Daulay and Singh 1982). Studies revealed that alachlor, trifluralin and nitrofen are found effective in controlling weeds and give yield at par to mechanical weeding (Daulay and Singh 1982). Basalin is an effective herbicide for controlling weeds when incorporated in loamy sand at 1 kg active ingredients/ha (Kumar et al. 1996) and at 1.5 kg active ingredients/ha in sandy loam soil (Yadav et al. 1997c). The use of various herbicides requires proper attention such as soil type, moisture, time and method of application, environmental condition and interaction of these herbicides on disease incidence in the crop before their use.

4.16 Harvesting and Threshing

The varieties of clusterbean shown for seed purposes normally mature within 90–100 days depending upon type of varieties, rainfall, soil type and distribution of rainfall. The crop sown for fodder and green manure purpose should be harvested at 50 % flowering stage to get maximum quality benefits. While crop sown for seed purposes may be harvested at the maturity of crop depending upon the varieties and weather conditions. Harvesting is preferred in the morning to avoid seed shedding. The crop is harvested with a sickle. The harvested crop is spread and after proper drying, trampled over by bullocks or tractor to thresh out the seed. After threshing, winnowing is done to separate seed and husk. Nowadays threshing machines are available for separation of seeds. The seeds stored at higher temperatures and higher relative humidity may lead to complete loss of viability (Doijode 1989). After threshing, farmers generally keep a part of their produce for sowing in the next season. It has been reported that the quality of farmers' saved seed is not up to the mark in terms of physical as well as genetical purity and its productivity is low as compared to quality seed (Arora et al. 1998). Seed quality of samples collected from farmers' saved seed, public and private seed sectors revealed that the quality of the seeds produced by public and private seed organizations was good and met the Indian minimum seed certification standards (IMSCS) while the quality in respect of farmers saved seed was marginally lower than the IMSCS in terms of germination, physical purity and moisture content (Kumar et al. 2012). Efforts must be made to promote the use of high quality certified seed amongst the farmers for good germination, good crop stand and higher productivity.

4.17 Clusterbean Varieties Developed in India

Clusterbean varieties grown in South India are vegetable types while those in north-west India are grown for seeds. There are giant and dwarf type varieties in clusterbean. Vegetable types are mostly dwarf types with smooth appearance and fodder types are mostly hairy. List of notified varieties of clusterbean, gazette notification number and date of notification are given in Table 4.3.

The general characteristics, agronomic features and reaction to diseases and pests of the notified varieties (dacnet.nic.in; agropedia.iitk.ac.in) are described as under:

Durgajay: The variety was developed from single plant selection in 1978. It is recommended for cultivation in Rajasthan for fodder as well as seed. The maturity period of the variety ranges from 110 to 115 days. It yields 2,7000 kg/ha green fodder and 1260 kg/ha seed.

Durgapura safed: This variety is recommended for cultivation in Rajasthan for green fodder and seeds and is suitable for late sown conditions. The maturity period of the variety ranges from 110 to 115 days and the average grain yield ranges from 1200 to 1500 kg/ha and green fodder yield is 25,000 kg/ha.

Table 4.3   Notified varieties of clusterbean

SN	Variety name	Gazette notification no.	Notified date
1.	HG-884	S.O. 733(E)	01-04-2010
2.	HG-2-20	S.O. 733(E)	01-04-2010
3.	HG-870	S.O. 733(E)	01-04-2010
4.	Rajasthan Guar-1038 (RGC-1038)	S.O. 449(E)	11-02-2009
5.	Gaur Lathi (RGC-1066)	1703(E)	05-10-2007
6.	Gaur Uday (RGC-1055)	1703(E)	05-10-2007
7.	Guar Kranti (RGC-1031)	1566(E)	05-11-2005
8.	Surya (RGM-112)	122(E)	02-02-2005
9.	HG-563	642(E)	31-05-2004
10.	RGC-1017	937(E)	04-09-2002
11.	Kanchan Bahar (M-83)	1135(E)	15-11-2001
12.	Bundel Guar-3 (IGFRI-1019-1)	1050(E)	26-10-1999
13.	RGC-1002	1050(E)	26-10-1999
14.	RGC-986	425(E)	08-06-1999
15.	RGC-1003	425(E)	08-06-1999
16.	Haryana Guar-365	401(E)	15-05-1998
17.	Bundel Guar-2	408(E)	04-05-1995
18.	Rajasthan Guar-1 (RGC-471)	408(E)	04-05-1995
19.	Bundel Guar-1 (IGFRI-212-1)	615(E)	17-08-1993
20.	RGC-936	793(E)	22-11-1991
21.	Gujarat Guar-1 (GAUG-34)	793(E)	22-11-1991
22.	RGC-197	386(E)	15-05-1990
23.	HFG-156	10(E)	01-01-1988
24.	HG-258	10(E)	01-01-1988
25.	Maru Guar	867(E)	26-11-1986
26.	Pusa Navbahar	258(E)	14-05-1986
27.	Durgabahar (RGC-955)	832(E)	18-11-1985
28.	HG-182	596(E)	13-08-1984
29.	Ageta Guara-112	499(E)	08-07-1983
30.	Guara-80	499(E)	08-07-1983
31.	HFG-11	19(E)	14-01-1982
32.	HFG-119	19(E)	14-01-1982
33.	HG-75	19(E)	14-01-1982
34.	Durgajay	470	19-02-1980
35.	Durgapura Safed	470	19-02-1980
36.	Agaita Guara-111	13	19-12-1978
37.	FS-277	786	02-02-1976
38.	Type-1		01-01-1968
39.	Type-2		01-01-1968
40.	Maru Guar (2470/12)		
41.	Guar Karan (RGC-1038)		

Source www.seednet.gov.in

Agaita guara-111: The variety was developed by PAU, Ludhiana in 1980. It is recommended for cultivation in all clusterbean growing areas of Punjab. The maturity period of the variety ranges from 110 to 111 days. It provides 23,000 kg/ha of green fodder and 4400 kg/ha dry fodder.

Agaita guara-112: The variety is an early maturing type and is recommended for cultivation in all clusterbean growing areas of Punjab state. The variety was developed in 1980 by PAU, Ludhiana. It provides 30,000 kg/ha of green fodder, 6400 kg/ha dry fodder and 1600 kg/ha of grain.

FS-277: The variety was developed by CCS HAU, Hisar and is an erect and unbranched variety recommended for cultivation in entire clusterbean growing areas of the country. It is well adapted to the states of Punjab and Haryana. The average grain yield is 2800 kg/ha and average green fodder yield is 21,100 kg/ha under timely sown conditions. The variety is drought tolerant and susceptible to bacterial blight and moderately resistant to Alternaria leaf spot.

HFG-119: The variety was developed by CCS HAU, Hisar in 1981 and is especially suited for fodder purposes. It is drought tolerant, non-shattering and resistant to Alternaria leaf spot and moderately resistant to bacterial blight. No serious pests in Haryana are reported on this variety. It is notified for cultivation in the entire clusterbean growing area of the country. The maturity period of the variety ranges from 130 to 135 days providing 25,000–30,000 green and 5000–6000 kg/ha of dry fodder.

HG-75: The variety was developed by CCS HAU, Hisar in 1981. It is recommended for cultivation in all clusterbean growing areas of the country for seed production. The maturity period of the variety ranges from 100 to 130 days. The variety yields 25,000 kg/ha green fodder and 1500–2000 kg/ha seed. The variety possesses field resistance to Alternaria leaf spot, bacterial blight and the major insect pests of the crop.

Guara-80: The variety was developed in 1982 by PAU, Ludhiana. It is recommended for cultivation in north-western zone of the country and is suitable for arid and humid areas. The maturity period of the variety ranges from 125 to 160 days and seed yield ranges from 1450 to 1500 kg/ha. It produces 26,800 kg/ha green fodder and 4660 kg/ha dry matter. The variety is resistant to water lodging and shattering. It is suitable for normal and late sown under both irrigated and rainfed conditions. 18–20 kg seed is required for sowing in one acre of area. The variety is suitable for intercropping with pearl millet and sorghum. The variety is comparatively more resistant to the diseases caused by *Xanthomonas cyamopsidis* and *Alternaria cyamopsidis*.

HG-182: The variety was developed by CCS HAU, Hisar in 1981. The maturity period of the variety ranges from 110 to 125 days and the seed yield ranges from 1500 to 1800 kg/ha. The variety is drought tolerant and possesses field resistance to most of the diseases and pests of the crop.

Maru guar (2470/12): The variety was developed by CAZRI, Jodhpur. The variety is dual type and is suitable for cultivation in western Rajasthan. It yields 22,500 kg/ha green fodder and 950 kg/ha seed. The maturity period of the variety ranges from 97 to 100 days and the seed yield ranges from 684 to 899 kg/ha. The variety is resistant against Alternaria and bacterial blight disease.

HFG-156: The variety was developed by CCS HAU, Hisar during 1986 for cultivation in Haryana. It is a tall, branched variety yielding 35,000 kg/ha green fodder. The maturity period of the variety ranges from 70 to 120 days and the average seed yield ranges from 500 to 600 kg/ha.

Bundel guar-1 (IGFRI-212-1): It was developed by Indian Grassland and Fodder Research Institute (IGFRI), Jhansi in 1991. The variety was recommended for cultivation in the entire clusterbean growing areas and is most suitable for cultivation in arid and semi-arid zones of the country under moderate to low rainfall conditions. The maturity period of the variety ranges from 120 to 135 days and the average yield ranges from 1200 to 1300 kg/ha. It provides 35,000 kg/h green fodder and 6500 kg/ha of dry fodder with a protein yield of 1150 kg/ha. The variety provides nutritive fodder within 50–55 days. This variety is moderately resistant to leaf blight under epiphytotic field conditions. It is resistant to water lodging, responsive to fertilizers, drought tolerant and has a non-shattering character. The variety has exhibited moderate resistance to bacterial blight and resistant to major pests of the crop.

Bundel guar-2 (IGFRI-2395-2): The variety was developed by IGFRI, Jhansi in 1993. The variety was recommended for cultivation in the entire clusterbean growing areas and is most suitable for cultivation in arid and semi-arid zones of the country under moderate to low rainfall conditions. Its green fodder, dry fodder and crude protein yield range between 25,000–30,000, 5000–6000 and 120–150 kg/ha, respectively. The variety has good palatability to the cattle with about 74 % digestibility on dry matter basis. It is superior in grain and gum production. The variety is highly responsive to fertilizers and moderately resistant to bacterial blight and has shown tolerance to lodging, drought and shattering. It has been released and notified as the first dual type (forage and grain) of clusterbean variety. The maturity period of the variety ranges from 120 to 135 days and the average yield ranges from 2500 to 2600 kg/ha. It is moderately resistant to bacterial blight and resistant to major pests of the crop.

Bundel guar-3 (IGFRI-1019-1): This variety was developed by IGFRI, Jhansi in 1997. The variety has been released and notified for general cultivation in entire clusterbean growing area of India as forage and grain type and is most suitable for cultivation in arid and semi-arid zones of the country with moderate to low rainfall conditions. The variety is moderately resistant to bacterial blight and powdery mildew, responsive to fertilizers, highly tolerant to shattering and reasonably resistant to drought situations. The maturity period of the variety ranges from 50 to 55 days and the average green fodder yield ranges from 35,000 to 40,000 kg/ha.

Guar kranti (RGC-1031): The variety was developed by ARS, Durgapura, Jaipur, RAU, Bikaner in 2005. The variety has wider leaves, whitish seed and is suitable for sowing under irrigated conditions. The variety is recommended for rainfed, well-drained and sandy loam soils and is suitable for cultivation in Rajasthan state during rainy season. It gave high grain and fodder yields under irrigated conditions. The maturity period of the variety ranges from 90 to 100 days and the average yield ranges from 1500 to 1600 kg/ha. The variety yields 34,000 kg/ha green fodder and 1460 kg/ha seed. The variety is resistant to lodging, shattering and responsive to 20 kg N and 40 kg P_2O_5/ha with 30 cm of row spacing and 20 kg seed is sufficient for sowing in one ha of area. The variety is suitable for late sown, rainfed conditions and is tolerant to drought and diseases, viz. bacterial blight, Alternaria blight, root rot, wilt, other diseases.

HG-884: The variety was developed by CCS HAU, Hisar for cultivation in Haryana, Uttar Pradesh, Madhya Pradesh, Gujarat and Rajasthan under timely sowing, normal fertility and rainfed conditions and was released in year 2008. The variety is resistant to bacterial blight and moderately resistant to Alternaria blight and root rot disease and shows good level of resistance against leafhopper and aphid. It is a variety with high gum content (30–31 %) and viscosity (3000–3500 cP). The medium maturity of the variety is 110 days and ranges from 100 to 110 days, whereas, the average yield is 1500 kg/ha ranging from 1400 to 1500 kg/ha. To get maximum production, the sowing of this variety should be done during the last week of June to first week of July at 45 cm row spacing using 15 kg seed/ha fertilized with 20 kg N and 40 kg P_2O_5/ha.

HG-2-20: The variety was recommended for cultivation in the states of Uttar Pradesh, Haryana, Gujarat and Rajasthan under timely sowing, normal fertility and rainfed conditions. The maturity period of the variety ranges from 90 to 100 days and yields 1000–1600 kg/ha. The variety was developed in 2008 by CCS HAU, Hisar. To get maximum production from this variety, sowing should be done during the last week of June to first week of July at 45 cm row spacing using 15 kg seed/ha, fertilized with 20 kg N + 40 kg P_2O_5/ha. The variety has exhibited field tolerance to bacterial blight, moderately resistance to Alternaria blight and root rot and no major problem of insect pest was recorded with this variety. The variety is suitable for sowing with wider spacing, irrigated conditions/high rainfall regions.

HG-870: The variety was released in year 2008 by CCS HAU, Hisar and was recommended for cultivation in coastal saline areas. To get maximum production from this variety, sowing should be done during the last week of June at 45 cm row spacing using 15 kg seed/ha and should be fertilized with 20 kg N + 40 kg P_2O_5/ha. The variety has exhibited field tolerance to bacterial blight, moderately resistance to Alternaria blight and root rot and it has showed good level of resistance against leafhopper and there was no attack of aphids.

Rajasthan guar-1038 (RGC-1038): The variety was released in 2007 for cultivation in the states of Haryana, Maharashtra, Rajasthan, Madhya Pradesh, Uttar Pradesh and Gujarat during rainy and winter seasons under rainfed conditions. The maturity period of the variety ranges from 101 to 105 days and the average yield is of 1200–1500 kg/ha. The variety was developed by ARS, Durgapura, Jaipur, RAU, Bikaner. The variety has higher grain and fodder yield capacity at 30 cm spacing using 20 kg seed/ha and has good growth responses with 20 kg N/ha and 40 kg P_2O_5/ha. It is suitable for late sown, rainy and summer conditions and is resistant to lodging, shattering. The variety is drought tolerant and has wider adaptability to rainfed conditions at Nagaur and Pali districts of western Rajasthan. It is somewhat photo insensitive in nature therefore, suitable to summer season with wider row spacing.

Guar lathi (RGC-1066): The variety was developed by ARS, Durgapura, Jaipur, RAU, Bikaner during 2007. The variety was recommended for cultivation in rainfed and well-drained soil conditions of Rajasthan state. The maturity period of the variety ranges from 97 to 105 days and the average yield ranges from 1032 to 1500 kg/ha. The variety is tall, single stem type with brisk podding behaviour and is suitable to mechanical harvesting in canal command areas with close planting. The variety has exhibited moderate resistance to bacterial and Alternaria blight, root rot and wilt.

Guar uday (RGC-1055): The variety was developed by ARS, Durgapura, Jaipur, RAU, Bikaner during 2007. It is recommended for cultivation in rainfed and well-drained soil conditions. The maturity period of the variety ranges from 96 to 106 days and the average yield ranges from 1096 to 2880 kg/ha.

Surya (RGM-112): The variety was developed by ARS, Mandore, Jodhpur, RAU, Bikaner in 2003. It is suggested for cultivation under rainfed condition for all the clusterbean growing areas of the country where erratic and low rainfall occurs frequently and recommended especially in drier areas of the north-western parts of India. The maturity period of the variety ranges from 85 to 99 days and the average seed yield ranges from 1000 to 1300 kg/ha. The variety is drought tolerant and moderately resistant to bacterial blight, jassids and highly resistant to lodging.

HG-563: The variety was developed by CCS HAU, Hisar in 2003. It is recommended for cultivation in the normal monsoon season. The maturity period of the variety ranges from 85 to 100 days and the average seed yield ranges from 807 to 855 kg/ha. It is an early maturing, medium-statured and lodging resistant variety, hence more suitable to intensive cropping system with narrow spacing. The variety has heavy podding behaviour, improved in gum content, viscosity profile (4050 cP) and well suited to low rainfall zones. The seeds are of grey colour and have about 30 % of gum content. It is tolerant to bacterial blight and no major problems of any pest are recorded.

RGC-1017: The variety was developed from ARS, Durgapura, Jaipur in 2000 and recommended for cultivation in Rajasthan, Gujarat, Haryana under rainfed conditions as well as drain soil. The maturity period of the variety ranges from 92 to 99 days and the average seed yield ranges from 1000 to 1400 kg/ha. No major disease and pest were recorded with this variety.

Kanchan bahar (M-83): The variety was developed from ARS, Durgapura, Jaipur, RAU, Bikaner in 1999. It is recommended for the cultivation in all the vegetable growing areas of Rajasthan. The maturity period of the variety ranges from 85 to 90 days and the average seed yield ranges from 7500 to 7800 kg/ha. The matured pods of the variety is resistant to shattering and the variety is suitable for sowing in the summer and rainy seasons. The variety is tolerant to the major diseases, viz. bacterial blight, powdery mildew, Alternaria leaf spot, rhizoctonia rot and pests under field conditions.

RGC-1002: The variety was developed from ARS, Durgapura, Jaipur, RAU, Bikaner in 1997. It is recommended for the cultivation in all the clusterbean growing areas of India under well-drained soil conditions. It is well adapted to Sikar and Jaipur districts of Rajasthan with average rainfall of 375–650 mm. The maturity period of the variety ranges from 80 to 90 days and the average seed yield ranges from 1000 to 1300 kg/ha. No major disease and pest were recorded with this variety.

RGC-986: The variety was developed from ARS, Durgapura, Jaipur, RAU, Bikaner in 1997 for the cultivation in all the clusterbean growing areas of India under rainfed and well-drained soil conditions. The maturity period of the variety ranges from 110 to 115 days and the average seed yield ranges from 500 to 600 kg/ha. No major disease and pest were recorded but the variety requires better management practices. It is dual purpose, tall growing variety suitable for canal command areas.

Haryana guar-365: The variety was developed by the Department of Plant Breeding, CCS HAU, Hisar in 1997 for cultivation in Haryana state under rainfed conditions. This medium height, spreading-type variety has high viscosity and is one of the ruling varieties of Haryana. It is also grown in Gujarat and Rajasthan. The maturity period of the variety ranges from 90 to 95 days and the average seed yield is 1000–1200 kg/ha. The variety is suitable for early as well as late sown conditions and is resistant to lodging and shattering. It is resistant to bacterial blight and tolerant to environmental stresses and major insect pests of the crop.

Rajasthan guar-1 (RGC-471): The variety was developed from ARS, Durgapura, Jaipur, RAU, Bikaner in 1993 for the cultivation in all the clusterbean growing areas of the country during rainy season and well-drained soils. The maturity period of the variety ranges from 100 to 120 days and the average seed yield ranges from 1300 to 1400 kg/ha. The variety is moderately resistant to bacterial blight.

RGC-936: The variety was developed from ARS, Durgapura, Jaipur, RAU, Bikaner in 1989 for the cultivation in the well-drained soils and drought prone areas of Rajasthan, Gujarat and Haryana during rainy season. The variety is drought hardy with light pink flowers and well suited for arid areas, viz. Churu, Bikaner with average rainfall of 170–200 mm. The maturity period of the variety ranges from 70 to 90 days and the average seed yield ranges from 900 to 1200 kg/ha. No pest was observed under field and store conditions.

Gujarat guar-1 (GAUG-34): The variety was developed from Gujarat Agricultural University, Sardar Krushi Nagar in 1989. It can be grown under low fertility and rainfed condition during rainy season. The variety is recommended for cultivation in all the clusterbean growing states of India. The maturity period of the variety ranges from 40 to 57 days and the average seed yield ranges from 1000 to 1300 kg/ha. The variety is resistant to bacterial blight and Alternaria leaf spot under field as well as epiphytotic conditions.

RGC-197: The variety was developed from ARS, Durgapura, Jaipur, RAU, Bikaner in 1988 and is recommended for the cultivation in all the clusterbean growing areas of the country. The maturity period of the variety ranges from 100 to 120 days and the average seed yield ranges from 1000 to 1800 kg/ha. The variety is well tolerant to drought conditions, resistant to lodging and mature pods are non-shattering. The variety is tolerant to the major diseases, viz. bacterial blight, powdery mildew, Alternaria leaf spot, rhizoctonia rot and no pests were observed under field and store conditions.

HG-258: The variety was developed by CCS HAU, Hisar in 1986 for cultivation in all the clusterbean growing areas of the country. The variety is branched with pubescent leaves having smooth margin, its pods are medium in length, the seed colour is white and round in shape. The maturity period of the variety ranges from 110 to 125 days and the average seed yield ranges from 500 to 600 kg/ha.

Pusa navbahar: The variety was developed by the National Bureau of Plant Genetic Resources, New Delhi in 1984. It is single stemmed, non-branched and photo insensitive in nature. The pods are 12–15 cm long, light green, soft with profuse bearing in branches. The maturity period of the variety ranges from 120 to 125 days and the average seed yield ranges from 500 to 600 kg/ha.

RGC-1003: The variety was developed from ARS, Durgapura, Jaipur, RAU, Bikaner in 1997 for the cultivation in the well-drained soils of all the clusterbean growing areas of the country under rainfed conditions. The maturity period of the variety ranges from 110 to 115 days and the average seed yield ranges from 500 to 600 kg/ha. No major disease and pest were recorded with this variety.

Guar karan (RGC-1038): The variety was developed by ARS, Durgapura, Jaipur, RAU, Bikaner and is recommended for cultivation in the North Indian states in well-drained sandy loam soil under rainfed conditions during both rainy and summer seasons for both grain and fodder yields. The maturity period of the variety ranges from 101 to 105 days and seed yield ranges from 1000 to 1100 kg/ha. The drought tolerant variety is resistant to lodging, shattering and responsive to 20 kg N and 40 kg P_2O_5 with 30 cm row spacing. The variety is suitable for late sown conditions in rainy and summer cultivations using 20 kg seeds/ha. The variety shows average reactions to major diseases, viz. bacterial blight, Alternaria blight, root rot, wilt and other diseases.

Type-1: The variety was developed in 1966. The maturity period of the variety ranges from 135 to 140 days and the average seed yield ranges from 500 to 600 kg/ha. The variety is suggested for sowing with row to row and plant to plant distance of 30 and 10^{-15} cm, respectively, with the seed rate of 12–15 kg seeds/ha.

Type-2: The variety was developed in 1966. The maturity period of the variety ranges from 90 to 100 days and the average seed yield ranges from 500 to 600 kg/ha. The variety is recommended for sowing with row to row and plant to plant distance of 30 × 10^{-15} cm, respectively, with the seed rate of 12–15 kg seeds/ha.

Durgabahar (RGC-955): It is an unbranched variety having long smooth pods, white flowers and matures between 100–110 days and yields 7000–7500 kg green pods/ha. The variety was notified in 1985 for cultivation all over India for vegetable purpose. The first picking of the crop can be harvested after 45–50 days of sowing. The plants are erect, photosensitive, single stemmed and long pods are born in clusters. Durgabahar gives higher green pod yield under rainfed conditions.

RGC-1033: The plants of the variety are branched with non-serrated leaves and pink flowers. Seeds contain 34–35 % gum. The variety is recommended for cultivation in all the clusterbean growing areas of Rajasthan. The maturity period of the variety ranges from 95 to 106 days.

4.18 Clusterbean Production in India

The production of clusterbean in India is mainly confined to arid zones of Rajasthan and parts of Gujarat, Haryana and Punjab. In Rajasthan, it is mainly cultivated under rainfed conditions. The area, production and yield of the crop are inconsistent due to over reliance of the crop on weather. India accounts for more than three-fourth of the total world clusterbean production. The other major producers of clusterbean are Pakistan, USA, South Africa, Malawi, Zaire and Sudan.

The crop is considered generally for marginalized land but due to significantly higher prices owing to the demand of clusterbean seed/gum in international market during the recent time, it has got a lot of curiosity among farmers and helped in the expansion of the crop to the non-conventional regions and seasons. The crop is now being cultivated in dry tracts of Madhya Pradesh, Chhattisgarh, Andhra Pradesh, Karnataka, Tamil Nadu and other parts during rainy and summer seasons on variety of soils, viz. black, deep, clay, sandy, red. Seeing the demand and interest, region and season-based production techniques of the crop should be developed. It is also required for bridging the gaps between yields obtained on the experimental farm fields and at farmers' field.

Rajasthan is the largest clusterbean-producing state in India followed by Haryana and Gujarat and small contributions from the states of Uttar Pradesh, Punjab and Madhya Pradesh. The statistical data revealed that 95 % of the clusterbean production in India is coming mainly from Rajasthan and Haryana, whereas, Rajasthan has three-fourth contribution alone (Anonymous 2014). About 90 % area under the crop is consistently being contributed by Rajasthan however the production of the crop is comparatively low in this state showing great variations. In addition to Rajasthan, the crop is also cultivated in Haryana, Gujarat, Uttar Pradesh and Punjab. Haryana and Gujarat have almost similar area under clusterbean cultivation but the productivity of the crop in Haryana is significantly higher in the recent years. The high yielding and short duration varieties, viz. HG-365 and HG-563 and its extensive use by farmers have supported higher productivity in the state (NRAA 2014). Increasing area under cultivation of the crop in Haryana will help in boosting the overall production of clusterbean from India.

Clusterbean is grown in almost all the districts of Rajasthan except Baran, Jhalawar and Kota but about 97 % of the area is confined to 13 districts, namely Barmer, Bikaner, Churu, Hanumangarh, Jaipur, Jaisalmer, Jalore, Jhunjhunu, Jodhpur, Nagaur, Pali, Sikar and Sri Ganganagar which have more than 1 % clusterbean area and produce 94 % of the crop in the state (NRAA 2014). Out of these ten districts, i.e. Barmer, Bikaner, Churu, Hanumangarh, Jaisalmer, Jhunjhunu, Jodhpur, Nagaur, Sikar and Sri Ganganagar occupy about 92 % of the area and contribute about 88 % of the production with productivity level of 399.53 kg/ha.

The farmers in Ananthapur, Guntur, Karimnagar, Karnool, Nellor, Prakasam and Rangareddi districts of Andhra Pradesh also have started cultivation of clusterbean. In addition to Andhra Pradesh, clusterbean cultivation has also been introduced in Karnataka, Maharashtra and Tamil Nadu. Similarly, the crop has also

Table 4.4 Area, production and yield of clusterbean in India

Year	Rajasthan			Haryana			All India level		
	Area	Production	Yield	Area	Production	Yield	Area	Production	Yield
1970–71	1171	431	370	94.1	68	720	1465	651	440
1971–72	1197	244	200	106.1	53.9	510	1539	517	340
1972–73	774	95	120	141.2	97.8	690	1184	460	390
1973–74	1338	415	310	99.2	75.9	770	1710	760	440
1974–75	931	120	130	139.7	83.8	600	1221	307	250
1975–76	2294	476	210	173.6	130.4	750	2722	747	270
1976–77	2405	855	360	181.9	128.2	700	2812	1110	400
1977–78	1957	687	350	177.1	166.8	940	2367	965	410
1978–79	2121	750	360	203	186.6	920	2587	1069	410
1979–80	1435	144	100	179.9	124.9	690	1829	376	210
1980–81	1965	315	160	250.1	143.3	570	2402	568	240
1981–82	1669	313	190	299.7	187.3	630	2172	611	280
1982–83	1996	282	140	297.9	193	650	2451	610	250
1983–84	2208	663	300	327.2	208.5	640	2723	1015	370
1984–85	2038	360	180	312.8	191.8	610	2516	702	280
1985–86	1803	142.1	80	274.3	162.3	590	2195	390	180
1986–87	1751.4	131.8	80	247.5	154	620	2102	341	160
1987–88	947.3	29.2	30	120.7	85.5	710	1102	144	130
1988–89	2105	573	270	330	208	630	2585	901	350
1989–90	1971	445	230	182	131	720	2268	654	290
1990–91	2090	946	453	204	148	725	2403	1176	489
1991–92	1559	204	131	131	94	718	1765	346	196
1992–93	1882	583	310	155	93	600	2179	797	366
1993–94	1897	287	151	161	119	739	2101	490	233
1994–95	1959	708	361	156	117	750	2302	939	408
1995–96	1775	274	155	136	104	765	2213	900	407
1996–97	1819	740	407	127	104	819	2125	886	417
1997–98	1985	734	370	137	109	796	2301	963	418
1998–99	1612	320	198	127	82	646	1922	489	254
1999–00	2649	232	87	133	88	662	2934	375	128
2000–01	3056	481	157	148	102	689	3497	659	188
2001–02	2413	763	316	196	127	648	2903	1090	375
2002–03	557	28	50	205	91	444	974	203	208
2003–04	2278	1163	511	269	117	435	2854	1513	530
2004–05	1944	368	189	217	254	1171	2867	903	315
2005–06	2445	593	243	270	289	1070	2956	1059	358
2006–07	2808	658	234	295	334	1132	3352	1100	328
2007–08	2310	622	269	300	408	1200	2849	1262	443
2008–09	3316	1261	380	370	602	1627	3863	1936	501
2009–10	2581	201	78	252	329	1305	2996	595	199

(continued)

Table 4.4   (continued)

Year	Rajasthan			Haryana			All India level		
	Area	Production	Yield	Area	Production	Yield	Area	Production	Yield
2010–11	3001	1546	515	256	333	1300	3382	1965	581
2011–12	3000	1847	616	215	290	1350	3444	2218	644
2012–13	4526	2023	447				5152	2461	478
2013–14	4924	2201	512				5603	2715	485
2014–15							4255	2415	567

Area in '000 ha, Production in '000 tonnes and yield in kg/ha
Source Apeda.gov.in

been cultivated in Allahabad, Kannauj, Kaushambi, Mirzapur and Sonbhadra districts of Uttar Pradesh and seven districts of Bundelkhand.

There was a wide variation in the production of clusterbean at all India level ranging from 0.2 to 2.7 million metric tonnes (Table 4.4). Due to severe drought conditions during 2002–2003 only 0.2 million metric tonnes of clusterbean could be produced whereas, it was highest during 2013–2014. Similar trend in the area under the crop was also recorded. It was the lowest during 2002–2003 and highest during 2013–2014. The yield of the crop was highest (489 kg/ha) during 1990–1991 and was lowest (128 kg/ha) during 1999–2000 due to the over dependence of the crop on monsoon precipitation. The greater fluctuation in the area and production of clusterbean are evident from the fact that a harvest of about 2.2 million tonnes could be obtained out of 3.4 million ha area during 2011–2012, but it could be culminated into higher productions during 2009–2010 out of about 3 million ha area.

The area cultivated within India diminished during 2014, because most of the clusterbean is grown under rainfed conditions. A large part of the first sowing was destroyed due to drought in rainfed growing belt of the crop. This deficit resulted in increase in the cost of the crop (Agriculture Information.com 2014). The price of seed is an important factor for allocation of area under clusterbean cultivation. Besides this, relative profitability, traditional cropping pattern, availability of resource, irrigation and market demand also influence the farmers' decision for sowing of the crop. In general due to resource constraints, the farmers growing clusterbean under rainfed conditions do not shift to other crops; whereas the cultivation moderately shifts to other crops while clusterbean is grown under irrigated condition.

References

Agriculture Information.com (2014) Guar seed (Guar Gum). Cultivation details and other information. http://www.agricultureinformation.com/forums/consultancy-services/119615-guar-seed-guar-gum-cultivation-details-other-information-4.html
Angus JF, Cunningham RBM, Moncur W et al (1981) Phasic development in field crops-thermal response in the seedling phase. Field Crop Res 3:365–378

Anonymous (2013) Annual progress report. Central Arid Zone Research Institute, Jodhpur (Rajasthan), India, p 40

Anonymous (2014) Global agricultural information network report number IN 4035. USDA Foreign Agricultural Services, p 41

Arora RN, Ram H, Tyagi CS et al (1998) Comparative performance of farmers saved seed vis-à-vis quality seed for yield and seed quality in guar (Cyamopsis tetragonoloba (L.) Taub). Forage Res 24(3):159–162

Atwal AS, Sidhu GS (1964) Legumes in the nitrogen economy of soil: fixation and excretion of nitrogen by Indian legumes under controlled conditions in sand culture. Indian J Agric Sci 34(3):139–145

Bains DS, Dhillon AS (1977) The influence of sowing dates and row spacing patterns on the performance of two varieties of clusterbean (Cyamopsis tetragonoloba). J Res Punjab Agric Univ 14(2):157–161

Bhadoria RBS, Jain PC (2005) Crop-weed competition studies on clusterbean under rainfed condition. Forage Res 31(2):97–98

Bhadoria RBS, Kushwaha HS (1995) Response of clusterbean to row spacing and phosphorus application. Forage Res 21(3):155–157

Bhadoria RBS, Chauhan DVS, Bhadoria HS (1996) Effect of weed control on yield attributes and yield of clusterbean (Cyamopsis tetragonoloba). Indian J Agron 41(4):662–664

Bhati DS, Manohar SS (1989) Studies on the intercropping of pearl millet with legumes at two nitrogen levels. Haryana J Agron 5(1):9–12

Chaudhary BO, Singh VP (1976) Studies on variability, correlations and path analysis in guar. Forage Res 2(2):97–103

Chauhan GS, Dungarwal HS (1982) Production potential of maize (Zea mays L.) with different companion crops and their methods of planting. Indian J Agric Res 16:193–198

Chavan BL, Vedpathak MM, Pirgonde BR (2015) Effects of organic and chemical fertilizers on clusterbean (Cyamopsis tetragonolobus). Eur J Exp Biol 5(1):34–38

Daisy EK (1979) Food legumes. Tropical Products Institute, London

Datta KS, Dayal J (1988) Effect of salinity on germination and early seedling growth of guar (Cyamopsis tetragonoloba). Indian J Plant Physiol 31:357–363

Daulay HS, Singh KC (1982) Chemical weed control in green gram and clusterbean. Indian J Agric Sci 52(11):758–763

Daulay HS, Henry A, Bhati TK (1998) Compatibility of promising varieties of clusterbean for inter/mixed cropping in pearl millet in dryland of western Rajasthan. Forage Res 24(1):13–16

Dhukia RS, Singh KP (1988) Effect of sowing dates and irrigation levels on grain quality of clusterbean. Forage Res 14(1):23–27

Doijode SD (1989) Deteriorative changes in clusterbean seeds stored in different conditions. Vegetable Sci 16:89–92

Elsayed ME (1994) The influence of locality and genotype on quality aspects of faba bean (Vicia faba L.) cultivars. M.Sc. (Agric.) thesis, University of Khartoum, Sudan

FAO Ecocrop (2007) The crop environmental requirements database. Rome. http://ecocrop.fao.org/ecocrop/srv/en/cropView?id=830

Francois LF, Donovan TJ, Maas EV (1990) Salinity effects on emergence, vegetative growth and seed yield in guar. Agron J 82(3):587–592

Gill PS, Singh K (1981) Effect of fertilization on yield contributing characters and grain yield of clusterbean (Cyamopsis tetragonoloba L. Taub.) varieties. Haryana Agric University J Res 11(3):333–338

Hymowitz T, Matlock RS (1963) Guar in the United States. Oklahoma Agric Exp Station Tech Bull 611:1–34

Hymowitz T, Matlock RS (1964) Guar: seed, plant and population studies. Okla Agric Exp Station Tech Bull (B) 108:1–35

Kalavathi D, Ramamoorthy K (1992) A note on the effect of seed size on viability and vigour of seed in clusterbeans Cyamopsis tetragonoloba L. cultivar Pusa Navbhagar. Madras Agric J 79(9):530–532

Kumar V, Yadav BD, Agarwal SK (1996) Performance of fluchloralin and pendimethalin to control weeds in clusterbean under rainfed condition. Haryana J Agron 12(1):43–46

Kumar A, Jakhar SS, Dahiya OS et al (2012) Seed quality status of clusterbean seed produced by farmers' as well as public and private sectors in Haryana. Forage Res 38(3):184–185

Kumhar MK, Patel IC, Ali S (2012) Integrated nutrient management in clusterbean (*Cyamopsis tetragonoloba* L. Taubert). Legume Res 35(4):350–353

Madalgeri MS, Rao MM (1989) Effect of fertilizer levels and staggered removal by early produced fresh pod on the quality of Pusa Navbahar clusterbean (*Cyamopsis tetragonoloba*) seeds. Seeds Farms 15:7–10

Meena KC, Singh GD, Mundra SL (1991) Effect if phosphorus, micro-nutrients and irrigation on clusterbean. Indian J Agron 36(2):272–274

Musil AF (1946) The germination of guar (*Cyamopsis tetragonoloba* (L.) Taub). J Am Soc Agron 38:661–662

Nandwal AS, Dabas S, Bharti S et al (1990) Zinc effect on nitrogen fixation and clusterbean yield. Ann Arid Zone 29(2):99–103

NRAA (2014) Potential of rainfed guar (Clusterbeans) cultivation, processing and export in India. Policy paper no. 3 National Rainfed Area Authority, NASC Complex. DPS Marg, New Delhi, 109 p

Palankar GS, Malabasari TA (2014) Effect of major nutrient and picking stage on seed yield and quality of clusterbean (*Cyamopsis tetragonoloba* L. Taub). J Agric Allied Sci 3(4):8–12

Pathak R, Roy MM (2015) Climatic responses, environmental indices and interrelationships between qualitative and quantitative traits in clusterbean under arid conditions. Proc Natl Acad Sci India Sec B (Biol Sci) 85(1):147–154

Prajapati BH, Chaudhari GK, Parmar HD et al (2004) Optimum plot size for field experiments in clusterbean (*Cyamopsis tetragonoloba*). Forage Res 30(3):121–124

Ramawtar, Shivran AC, Yadav BL (2013) Effect of NP fertilizers, vermicompost and sulphur on growth, effect of NP fertilizers, vermicompost and sulphur on growth, yield and quality of clusterbean [*Cyamopsis tetragonoloba* (L.)] and their residual effect on grain yield of succeeding wheat [*Triticum aestivum* (L.)]. Legume Res 36(1):74–78

Rana DS, Yadav BD, Dhukia RS et al (1991) Effect of row spacing and intra row spacing on seed yield of clusterbean under late sown conditions. Guar Res Ann 7:54–56

Reddy AM, Reddy BS (2011) Effect of planting geometry and fertility level on growth and seed yield of clusterbean (*Cyamopsis tetragonoloba* (L.) under scarce rainfall zone of Andhra Pradesh. Legume Res 34(2):143–145

Reddy GS, Venkateswarlu S (1989) Effect of fertilizer and planting pattern in castor-clusterbean intercropping system. J Oilseed Res 6(2):300–307

Reddy DS, Nagre PK, Reddaiah K et al (2014) Effect of integrated nutrient management on growth, yield, yield attributing characters and quality characters in clusterbean [*Cyamopsis tetragonoloba* (L.) Taub.]. Ecoscan 6:329–332

Renugadevi J, Natarajan N, Srimathi P (2009) Influence of seed size on seed and seedling quality characteristics of cluster bean [*Cyamopsis tetragonoloba* (L.) Taub.]. Legume Res 32(4):301–303

Satyavathi P, Vanaja M, Gopala Krishna Reddy A et al (2014) Identification of suitable guar genotypes for summer season of semi-arid region. Int J Appl Biol Pharma Technol 5(4):71–73

Saxena A, Singh DV, Joshi NL (1997) Effect of tillage and cropping systems on soil moisture balance and pearl millet yield. J Agron Crop Sci 178:251–257

Sharma OP, Singh GD (2005) Effect of sulphur in conjunction with growth substances on productivity of clusterbean (*Cyamopsis tetragonoloba*) and their residual effect on barley (*Hordeum vulgare*). Indian J Agron 50(1):16–18

Sharma BD, Taneja KD, Kairon MS et al (1984) Effect of dates of sowing and row spacing on yield and quality of Clusterbean. Indian J Agron 29:555–559

Shekhawat PS, Rathore AS, Singh M (1996) Effect of source and level of sulphur on yield attributes and seed yield of clusterbean (*Cyamopsis tetragonoloba*) under rainfed conditions. Indian J Agron 41(2):340–342

Shivran AC, Knangarot SS, Shivran PL et al (1996) Response of clusterbean (*Cyamopsis tetragonoloba*) varieties to sulphur and phosphorus. Indian J Agron 41(2):340–342

Singh BP (1981) Effect of intercropping of guar in pearl millet. Guar NewsLetter 2:24–27

Singh SP, Ahuja KN (1990) Intercropping of grain sorghum with fodder legumes under dry land conditions in north Western India. Indian J Agron 35(3):287–296

Singh M, Joshi NL (1994) Effect of mixed and intercropping systems on dry matter and grain yields of component crops in arid environment. Ann Arid Zone 33:125–128

Singh RP, Kaushik MK (1997) Nitrogen economy in maize legume intercropping system. Ann Agric Res 8:105–109

Singh JP, Singh BP (1988) Intercropping of green gram and guar in castor under dry land condition. Indian J Agron 33:177–180

Singh RV, Singh RR (1989) Effect of nitrogen, phosphorus and seeding rates on growth, yield and quality of guar under rainfed conditions. Indian J Agron 34(1):53–56

Singh R, Gupta PC, Paroda RS et al (1976) In vivo studies on the nutritive value of cowpea and guar. Forage Res 2:36–40

Singh DP, Rathore DN, Singh H, Kumar V (1978) A note on crude protein and gum production of two varieties of guar (*Cyamopsis tetragonoloba* (L.) Taub.) as influenced by different seed rates and row spacing. Ann Arid Zone 17(3):329–331

Singh YP, Dahiya DJ, Kumar V et al (1993) Effect of nitrogen application on yield and uptake of nitrogen by different legume crops. Crop Res 6(3):394–400

Singh V, Sharma SK, Ram Deo et al (1998) Performance of different crop sequences under various irrigation levels. Indian J Agron 43(1):38–44

Stafford RE, Hymowitz T (1980) Guar. In: Fehr WR, Hadley HH (eds) Hybridization of Crop Plants. American Society of Agronomy, Crop Science Society of America, Madison, Wisconsin, pp 381–392

Takkar PN, Mann NN, Randhawa NS (1973) Major rabi and kharif crops respond to zinc. Indian farm 23(8):5–8

Taneja KD, Sharma BD, Gill PS (1981) Response of clusterbean to varying levels of phosphorus and bacterial culture. Guar Newsletter 2:20–23

Taneja KD, Bishnoi OP, Rao VUM et al (1995) Effect of environment on growth and yield of clusterbean. Crop Res 9(1):159–162

Tiwana US, Tiwana MS (1992) Effect of sowing time, seed rate and row spacing on the seed yield of guar under rainfed conditions. Forage Res 18(2):151–153

Tiwana US, Tiwana MS (1993) Effect of sowing dates, seed rate and spacing on the seed yield of guar (*Cyamopsis tetragonoloba* (L.) Taub.) under irrigated conditions. Forage Res 19(2):115–118

Tyagi CS, Paroda RS, Lodhi GP (1982) Seed production technology for guar. Indian Farm 32:7–10

Undersander DJ, Putnam DH, Kaminski AR et al (1991) Guar. In: University of Wisconsin Cooperative Extension Service, University of Minnesota Extension Service, Center for Alternative Plant and Animal Products. Alternative Field Crop Manual. http://www.hort.purdue.edu/newcrop/afcm/guar.html

Venkateswarlu S, Subramanian VB (1990) Productivity of some rainfed crops in sole and intercrops system. Indian J Agric Sci 60:106–109

Yadav RS (1998) Effect of weed removal in clusterbean under different rainfall situations in an arid region. J Agron Crop Sci 181(94):209–214

Yadav BD, Joon RK (1993) Studies on inter cropping of pearl millet in guar. Forage Res 19(3&4):306–309

Yadav BD, Agarwal SK, Arora SK (1989) Yield and quality of new cultivars of clusterbean as affected by row spacing and fertilizer application. Guar Res Ann 5:24–27

Yadav BD, Agarwal SK, Faroda AS et al (1990) Physiological basis of yield variation in clusterbean as affected by row spacing and fertilizer application. Forage Res 16(1):38–41

Yadav BD, Agarwal SK, Faroda AS (1991a) Dry matter accumulation and nutrient uptake in clusterbean as affected by row spacing and fertilizer application. Forage Res 17(1):39–44

Yadav BD, Joon RK, Lodhi GP et al (1991b) Effect of agro management practices on the seed yield of clusterbean. Guar Res Ann 7:30–33

Yadav BD, Joon RK, Sheoran RS (1992) Response of clusterbean to date of sowing and row spacing under rainfed conditions. Forage Res 18(2):157–159

Yadav VK, Yadav BD, Joshi UN (1993) Effect of weed control and fertilizer application on seed and gum yield of clusterbean under rainfed condition. Forage Res 19(3, 4):341–342

Yadav RP, Sharma RK, Shrivastava UK (1997a) Fertility management in pigeon pea (*Cajanus cajan*) based intercropping system under rainfed conditions. Indian J Agron 42(1):46–49

Yadav BD, Joon RK, Singh VP (1997b) Crop weed competition in clusterbean under rainfed conditions. Haryana J Agron 13(2):92–96

Yadav BD, Joon RK, Lodhi GP (1997c) Chemical weed control in clusterbean (*Cyamopsis tetragonoloba* (L.) Taub.). Forage Res 23(3, 4):189–192

Yadav BD, Joon, RK, Lodhi GP (1998) Chemical weed control in clusterbean (*Cyamopsis tetragonoloba* (L.) Taub.). Forage Res 23(3, 4):189–192

Yadav BK, Rawat US, Meena RH (2012) Influence of phosphorus and sulphur on yield and micronutrient uptake by clusterbean [*Cyamopsis tetragonoloba* (L.) Taub]. Legume Res 35(1):8–12

Yousif YH (1984) Guar agronomy. Shambat Research Station, Annual Report Soba Research Unit, pp 3–4

Zade VR, Patel VN, Zade NG (1990) Standardization of seed testing procedures for *Trigonella foenum-graecum* and *Cyamopsis tetragonolobus*. Ann Plant Physiol 4(2):182–185

Zheng GH, Gu Z, Xu BH (1980) A physiological study of germination in guar (*Cyamopsis tetragonoloba*). Acta Phytophysiol Sin 6:115–126

Chapter 5
Plant Protection

Abstract Clusterbean is susceptible to a number of diseases but bacterial blight, Alternaria leaf spot and powdery mildew are some of the serious diseases which appear almost every year and badly affect the production and productivity of the crop. Minimum variation between day and night temperatures, cloudy climate, less sunshine hours favour the attack of diseases. The development of varieties with inbuilt resistance to the diseases is the holistic and practical approach for solving this problem. A number of germplasms and improved lines have been screened for disease resistance potential. A detailed account of fungal, bacterial, viral diseases and major insect pests of the crop has been discussed.

5.1 Introduction

Plant protection measures have substantial importance in the overall crop production programmes for sustainable agriculture to limit the yield losses during the growing season and afterwards for quarantine purposes. In combination with other cultivation measures it helps to raise yields of the crop (Schut et al. 2014). The objective of plant protection is minimizing crop losses due to consequences of weeds, diseases, insect pests, nematodes, etc. Wide range of ecological, economic and socio-economic impacts is available for keeping harmful organisms below the economic threshold.

Clusterbean is susceptible to a number of diseases but bacterial blight, Alternaria leaf spot and powdery mildew are some of the serious diseases which appear almost every year and badly affect the production and productivity of the crop. Less difference between day and night temperatures, cloudy climate and less sunshine hours are favourable conditions for the attack of diseases in this crop (Pareek and Varma 2014). The holistic and practical approach for solving this problem is to develop varieties with inbuilt resistance to the diseases. A number of germplasms and improved lines have been screened for disease resistance potential (Gandhi et al. 1978b; Gupta 1997). The additive, dominance and epistatic

© Springer Science+Business Media Singapore 2015
R. Pathak, *Clusterbean: Physiology, Genetics and Cultivation*,
DOI 10.1007/978-981-287-907-3_5

gene interactions play important role towards disease resistance in clusterbean (Saharan et al. 2001), however, it is not understandable whether the genes responsible for these diseases are identical or not. Therefore, there is a need to identify the individual genes responsible for disease resistance using a precise genome map of the crop. Peroxidase enzyme and phenol are reported as the important players in imparting the resistance towards the major diseases of clusterbean (Kalaskar et al. 2014) and oxidative burst as an earliest observable aspect of a plants defence strategy (Wojtaszek 1997). Field under clusterbean during rainy season is always full with a number of weeds. Hoeing and weeding in the initial stages of plant growth reduces the weed–crop competition and increases soil aeration for bacterial growth.

5.2 Fungal Disease

Clusterbean is susceptible to various fungal diseases, viz. powdery mildew, anthracnose, dry root rot, Phymatotrichum root rot, Alternaria leaf spot, Cercospora leaf spot, Myrothecium leaf spot, Curvularia leaf spot, etc. A brief account of some of the fungal diseases is given below.

5.2.1 Alternaria Leaf Spot

The disease is caused by *Alternaria brassica* (Berk.) Sacc. (Butler 1918; Streets 1948), *A. cyamopsidis* Rangaswami and Rao (Rangaswami and Rao 1957) and *A. cyamopsidis* (Ell. and Ev.) Elliot (Sowell 1965). The disease is seed borne, affecting the crop across the country and was primarily reported in Bihar and Madras. It manifests initially on the leaf blade in the form of dark brown, round to irregular spots with the varying diameter of 2–10 mm and turns to greyish-brown in later stage of its advancement. In the early stage of infection, water-soaked spots, which appear on the leaf blade, later turn into greyish to dark brown with concentric zonation, demarked with light brown lines. In severe infections, several spots merge together and cover almost part of leaf blade resulting into chlorosis and simultaneously death of leaf. High yield loss (43–87 %) are recorded when the leaves are infected at seedling and reproduction stage (Sharma 1983). Excessive rainfall, humidity, acidic soil, temperature of 30 °C and 12 h cycles of light and darkness including early sown crop favour the incidence of Alternaria blight (Ayub et al. 1995; Saharan and Saharan 2004). Singh and Prasada (1972) reported 35 °C as the optimum temperature for mycelial growth of the fungus followed by 30, 25 and 20 °C. Yogendra et al. (1995) recorded maximum disease intensity at 25–31 °C temperature, 80 % relative humidity and high rainfall. The amount of neutral detergent fibre, acid detergent fibre, hemicellulose, cellulose, lignin, silica contents, polyphenol oxidase and peroxidase enzyme was observed significantly higher in the plant with the increase in the severity of Alternaria blight. However, crude protein and organic matter decreased with increase in disease severity (Saharan et al. 2001).

The disease may be controlled by sowing pathogen resistance varieties/cultivars (Gandhi et al. 1978b). Number of disease resistant variety, viz. GAUG 9406, GAUG 9005, GAUG 9003, GG 1 (Kumar 2005), etc. has been identified. Mathur et al. (1972) found zineb (0.2 %) and copper oxychloride as the most effective fungicide and recommended its two sprays at the interval of 15 days for controlling the disease. A combination of zineb and streptocycline has also been recommended in controlling Alternaria and bacterial blights.

5.2.2 Powdery Mildew

The first report of the disease in India is from Bombay and Madras (Butler 1918) followed by Pakistan (Khan 1958). The pathogens associated to cause powdery mildew are *Oidium taurica, Erysiphe polygoni* (Ratnam et al. 1985) and *Sphaerotheca fuliginea* (Chandra and Saxena 1990). The disease manifests only on leaves and perithecal stage of the fungus first appears on the dried leaves. The white powdery growth occurs on leaves, spreading to cover the stem and other plant parts. In severe cases, the entire plant dries up. Severity of disease may result defoliation of plant, weakening of the premature leaves and its death. The prolonged crop season, warm temperature (33 % or above), high humidity (more than 80 %) and bright sunshine are the most favourable conditions for disease development and causes considerable defoliation.

Two sprays of *N*-triaecyl-6, 6-dimethyl morpholine at the intervals of 15 days have satisfactorily controlled disease (Solanki and Singh 1976). In other studies 0.025 % Benlate (Reddy and Rao 1971), Tridemorph (Sharma 1984) and Bavistin (Ratnam et al. 1985) are found effective in controlling the different pathogens causing powdery mildew in clusterbean. Seed treatment with streptocycline and spraying a combination of streptocycline and dinocap is found to be effective in controlling powdery mildew and bacterial blight. Sowing of pathogen resistance varieties/cultivars treated with Thiram is the best practise to control the disease. Crop rotation should be followed to reduce the soil-borne inoculum of the fungus.

5.2.3 Dry Root Rot

Macrophomina phaseolina is the major causal agent of dry root rot in clusterbean (Prasad 1945) and infect the crop at any stage from pre-emergence to maturity resulting into seedling blight, ashy stem blight or dry root rot. The dry and warm climatic conditions favour the incidence of this disease. The disease may be seen in the grown plants with bronzing leaves on the branching and with dropping of the upper tender parts of the shoot. The infected plant may have a normal growth but does not bear clusters or pods and can be easily uprooted even with strong winds. Purkayastha et al. (2006) studied genetic diversity among 59 isolates of *M. phaseolina* causing dry rot on clusterbean and reported substantial variation in their aggressiveness. Chlorate-resistant isolates produced longer lesions in comparison with the chlorate-sensitive isolates. The majority of the *M. phaseolina* isolates from clusterbean were found to be chlorate-resistant and the isolates having chlorate-resistant phenotype were the most pathogenic on clusterbean.

Rotation with less susceptible crops like moth bean or pearl millet or keeping the land fallow reduces the population of the pathogen. Conservation of moisture with mulching of soil with pearl millet stover, low plant density and use of FYM has reduced the population density of *M. phaseolina* and simultaneously dry root rot incidence (Lodha 1996). Incorporation of zinc, copper or sulphur coupled with FYM also reduced the incidence of disease and increased the production of the crop (Akem and Lodha 1997). Sowing of pathogen resistant varieties is another natural way of controlling the disease.

5.2.4 Cercospora Leaf Spot

The disease is caused by *Cercospora psoraleae* (Ray) and has been reported first time from India (Vasudeva 1960). Some seeds develop a purple discoloration and dark effuse lesions and in many cases it occurs on the stem and leaves of the plant. The disease can be identified by purple to orange discoloration of the uppermost leaves. Seed infected by *Cercospora* may cause poor seed vigour and reduced germination (Lodha and Mawar 2002). Seeds with substantial amounts of discoloration should not be saved. Tillage and crop rotation are effective ways to reduce the survival of these fungi from season to season on infested crop residues.

5.2.5 Myrothecium Leaf Spot

This disease is caused by *Myrothecium roridum* and was reported by Arya (1956) at Jodhpur. Later, he isolated this pathogen from clusterbean and from seven other hosts with varied pathogenicity and reported that it is not host specific (Arya 1959). The disease appears like oil-soaked brownish spores of 10–12 cm in diameters and under favourable climatic conditions these spots become larger, and with the severity of infection, several spots merge together and the entire leaf becomes chlorotic and ultimately defoliate. The infected seeds lose their viability (Shivanna and Shetty 1986). Two or three sprays of 0.2 % Dithane Z-78 at the interval of 15 days may reduce the infection.

5.2.6 Curvularia Leaf Spot

Chand and Verma (1968) reported this disease from Hisar, India caused by pathogen *Curvularia lunata*. The infection is seen in the form of small circular to oval, scattered on the lamina of the leaf, however, no other plant part shows any symptom. With the advancement of the infection, the leaf shows a blighted appearance and finally drops down. Two sprays of aureofungin at the interval of 15 days are found to be effective in controlling the disease (Singh et al. 1974).

5.2.7 Anthracnose

The disease is caused by a fungus, viz. *Collelotrichum capsici* f. *cyamopsicola* and is characterized by black spots on the petiole of leaf and stem of the plant. The fungus is seed borne and symptoms may start at seedling stage. The most characteristic symptoms of the disease are black, sunken, crater-like cankers on the cotyledons, stem or pods. With the advancement of the diseases the stem splits parallel to the axis and along the ridge and the plant remained stunted and dry up (Kothari and Bhatnagar 1966).

Sowing of crop under wider spacing with healthy seeds reduces the incidence of disease. The seeds should be treated with hot water before sowing and excess irrigation should be avoided. Spray of fungicides like Dithane M-45 or Dithane Z-78 at 2 kg/1000 L of water may be done under one hectare area to control the disease.

5.2.8 Ascochyta Leaf Blight

Kernhamp and Hemerick (1953) found *Ascochyta imperfecta* as a pathogenic agent on clusterbean and reported small black lesions on the leaves and stem of the plant. The lesions get enlarged and the entire stem gets surrounded by a black sheet of fungal mycelium with the advancement of the disease. The floral parts and peduncles also get infected resulting in falling of the flower. Small dark brown to nearly black spots of various shape, colour and size develop on leaves. The sources of infection are dry plant residues and seeds with mycelium and pycnidia of the pathogen. Crop residue should not be left in the field and seeds should be treated with Thiram + Bavistine and carbofuran at 0.25 g/kg before sowing.

5.2.9 Fusarium Blight

This blight appears in the late-sown crop and is caused by *Fusarium moniliforme* (Sheild) Snyder and Hansen (Prasad and Desai 1952). The pathogen is soil borne and causes infection to the base of plants including roots and seedlings. The symptoms of the disease start on the stem with black streaks and progress towards both upper and lower parts of the plant. Poor emergence of seedlings may be the first symptom of this disease.

It has been observed that the crop sown under well drained sandy soils is less affected with this blight. Addition of organic manures also reduces the disease incidence. Seed treatment with Agrosan GN or Thiram or captan at 3 g/kg seed prevents the disease spread during germination and seedling emergence.

5.2.10 Rhizoctonia Blight

Rhizoctonia is a soil borne fungus that can rot the seeds prior to emergence from the soil. Showery weather coupled with high temperature is the most favourable condition for this blight. The causal agent of the disease is *Rhizoctonia solani* and causes blighting and rotting of leaves and young branches of clusterbean. Young seedlings develop brick red to brown, sunken lesions on the tap root and basal stem. The tips of branch and tap roots may rot off leaving reddish-brown stubs during severe infections. Crop rotation, use of disease-free seed and proper drainage of water in the field are the best cultural practises to limit the disease. Incorporation of mustard residue + vermicompost + summer irrigation gave 65 % disease control and 23.4 q/ha seed yield (Kumar et al. 2008).

Different plant extracts and bioagents were evaluated under in vitro and field conditions against *R. bataticola* causing root rot of clusterbean (Sharma et al. 2005). *Trichoderma harzianum* showed its supremacy in controlling root rot incidence under laboratory and pot conditions. While neem extract was adjudged best among all the treatments under field conditions and recorded least rot incidence and highest grain yield, followed by *Mirabilis jalapa* and Bavistin.

5.3 Bacterial Disease

5.3.1 Bacterial Blight

The disease is caused by *Xanthomonas cyamopsidis* and was recorded first time from Patna in India (Patel et al. 1953; Srivastava and Rao 1963). It has also been reported from USA (Orellana et al. 1965) and recently from China (Ren et al. 2014). Among the major diseases in clusterbean, it is the most destructive seed borne disease and limits productivity in all major clusterbean growing regions especially in irrigated lands and dry-upland environment depending upon the crop stage, variety and climatic conditions (Gandhi and Chand 1985). The congenial situations causing spread of this disease are scattered rains, high temperature and humidity. It appears as spot on the dorsal surface of the leaf with water-soaked or oily look and characterized with irregular, sunken, red to brown leaf spots surrounded by a narrow yellowish radiance. Several spots merge to form irregular patches and results in blight phase. The infection advances from leaf to the stem through petioles resulting in blackening and cracking of the stem. Singh and Swarup (1987) reported that the disease is favoured during rainy season. The crop of 4–5 weeks is most susceptible of the disease leading to their drying due to stem rotting and may cause yield losses from 32 to 68 %.

The increase or no change in peroxidase activity can be taken as a marker for resistance to bacterial leaf blight (Kalaskar et al. 2014). Number of workers has suggested various methods for control or eradication of the disease, viz. hot water seed treatment at 56 °C for 10 min (Singh and Swarup 1986), seed treatment with streptomycin (Lodha 1984). Besides, the seed treatment foliar spray of streptomycin at 5th and 7th week after sowing reduced the secondary spread of the disease (Lodha and Anantharam 1993). Number of bacterial blight resistant varieties/genotypes has been screened, viz. IC 922, PLG 74, PLG 191, PLG 453, HFG 380, W 522, 590, 647, 648, 676 (Gandhi et al. 1978a), G 85, 102, 126, 225 (Chahal et al. 1980), GP 380, GP 508, GP 509B (Karwasara and Chand 1982), HG 75 (Lodha 1984), GAUG 9406, GG 1, RGC 1027 (Kumar 2005), etc. Genotypes, viz. GG-2, HG-75 and HG-365 and GAUG-0522 may be utilized in the future breeding programmes for high yield and resistance to bacterial blight in clusterbean (Kalaskar et al. 2014). Glabrous varieties and some of pubescent-branched types show considerable tolerance to the disease (Srivastava and Rao 1963). Vijayanand et al. (1999) produced Polyclonal antibodies- *X. campestris* pv. *Campestris* and developed an ELISA protocol for the rapid, sensitive and specific detection of *X. campestris* pv. *Campestris* in clusterbean. Anil et al. (2012) studied the inheritance of bacterial leaf blight resistance using F_2 progenies of two crosses, i.e. HG 563 × PNB and HG 75 × PNB and observed the presence of inhibitory gene action for governing resistance to bacterial leaf blight disease in both the crosses.

5.3.2 Bacterial Leaf Spot

The disease is caused by *Pseudomonas cyamopsicola* and was reported first time from India (Rangaswami and Gowda 1963) followed by Australia and USA (Orellana et al. 1965). The disease appears with tiny water soaked like light brown, irregular shape lesions on the leaf which turned into large patches with the advancement of the disease. The severe infection results in poor plant growth and yield, due to defoliation of the plant. The disease spread very fast and can damage up to 80–90 % crops. Orellana et al. (1967a) reported another species of Pseudomonas (*P. syringae*) responsible for bacterial leaf spot producing round to irregular lesions of 1–5 mm in diameter with dark borders and light brown centres surrounded by chlorotic radiance.

Crop rotation is one of the important measures for this disease. Farmer may go for regular spray of bactericide like streptocycline at 5 gm/100 L water or plantomycine at 50gm/100 L or agromycine at 30 gm/100 L water.

5.4 Viral Diseases

It has been reported that clusterbean is the host of several viruses, viz. bean top necrosis virus (Le Beau 1947), rose mosaic vines (Fulton 1948), sun hemp mosaic virus (Raychaudhuri et al. 1962), strawberry mottle virus (Harris et al. 1986), tobacco necrosis virus (Fulton 1950), tobacco ring spot virus and top necrosis virus (Chestor and Copper 1944), tobacco mosaic virus (Sekar and Sulochana 1986) and tobacco streak virus (Fulton 1948).

5.4.1 Mosaic Virus

It is a sap-transmissible viral disease and was reported in clusterbean from Hyderabad (Vani et al. 1986). It is characterized as the typical mosaic mottling and crinkling of leaves resulting in moderate stunting and Phytomonas accumulation in leaves resulting drastic reduction in the plant growth and development (Begum and Madhusudan 1989).

5.4.2 Tobacco Ring Spot Virus (TRSv)

Top necrosis caused by TRSv is considered to be a lethal disease in clusterbean (Chestor and Copper 1944; Verma et al. 1962). The clusterbean selections are more susceptible to the strains of TRSv than to tobacco strains (Orellana 1967b). The wild species, viz. *Cyamopsis senegalensis* and *C. serrata* are found to be resistant to this viral infection. So hybridization of these wild species with culti-vated species might be useful.

5.5 Other Root Diseases

Several pathogens, viz. *R. solani, F. coeruleum, Sclerotium rolfsii* and *Neocomospora vasinfecta* are reported to be associated with the root rot disease of clusterbean (Smith 1945; Singh 1951). *Ozonium taxanum* var. *parasiticum* is reported to cause wilt and leaf spot (Mishra 1963) in clusterbean. The incidence of *S. rolfsii* increased with soil moisture stress and acidic conditions. A number of root rot resistant variety, viz. GAUG 9406, HGS 844, GG1 (Kumar 2005), etc. have been identified.

The intercropping of clusterbean with sorghum, cowpea and moth in different combination reduced the incidence of *R. solani* and *F. caerulcum* (Singh 1954).

It was also observed that increased FYM reduced the incidence of root rot and wilt in mature clusterbean plant (Singh and Singh 1957). Liming of soil and heavy manuring including addition of humus have also been found to reduce *S. rolfsii* (Mathur 1962). Zinc plays an important role against fungal invasion causing root diseases. Soil treatment with zinc (20 mg/kg) improves defence mechanism and provides resistance in plants against fungal diseases by enhancing activity of anti-oxidative enzymes (Wadhwa et al. 2014).

5.6 Seed Mycoflora

The clusterbean seed is reported to be contaminated with various internally (*Alternaria*, *Aspergillus* and *Fusarium*) or externally (*Chaetomium* and *Gloeosporium*) associated fungi and they cause seed rot and may be the initial source of inoculums for different diseases (Suryanarayana and Bhombe 1961). Some uncommon fungi, viz. *Helminthosporium*, *Gliiocladium*, *Phoma*, *Pleospora* and *Schizophylium* are also associated with the seeds of clusterbean (Singh and Chouhan 1973).

Agrosan GN, Captan, Zineb and Mancozeb used as seed dressers reduce seed borne infections and induced germination and vigour of seedlings. Aureofungin and PCNB are found more effective for the control of pre-emergence mortality.

5.7 Insect Pests of Clusterbean

A number of insect pests feed and grow on the leaves and pods of clusterbean right from sowing to harvesting of crop (Butani and Jotwani 1984). The severe attack of these pests adversely affects the seed yield of the crop. As such, clusterbean has no specific insect pests, but the crop grown in the vicinity of other crops becomes infested by various pests such as aphids (*Aphis medicaginis* Koch); blossom thrips [*Megalurothrips distalis* (Karny)] and *M. usitatus* (Bagn.); whitefly [*Bemisia tabaci* (Gennadius)]; stink bugs [*Coptosoma cribrarium* (F.)], *C. nazirae* Atk and [*Cyclopelta siccifolia* (Westw.)]; pea leaf miner [*Phytomyza horticola* Goureau (*atricornis auct.*)]; blossom midge (*Asphondylia* sp.); and weevils (species of *Alcidodes*, *Blosyrus*, *Cyrtozemia* and *Myllocerus*), especially *A. bubo* (F.), *A. collaris* (Pasc.) and *M. undecimpustulatus* var. *maculosus* Desbr. (*M. maculosus*) (Butani 1980). The crop is regularly damaged by jassid (*Empoasca kerri* Pruthi), whitefly (*B. tabaci* Gennadius) and thrips (*Megaleurothrips distalis* Karny) like sucking insect pests during cropping seasons in Gujarat (Butani 1979). An account of major insect pests associated with clusterbean is summarized as under:

5.7.1 Whitefly

Acaudaleyrodes rachipora: Patel et al. (2011) reported *A. rachipora* as one of the major sucking whitefly of this crop and found that it sucks the cell sap from the undersurface of the leaves and damage the leaves by causing yellowing and shedding and resulting into loss of vitality of the plant. Begomovirus is transmitted by whitefly that seriously infect clusterbean crop (Kumar et al. 2010). The virus-infected plants show curling, shortening and malformation of leaves and shortening of internodes and stems with overall stunting of plant. The infected plant shows high mortality rate and short life span.

Dichomeris ianthus: The pest commonly known as leaf perforator attacks during the early stage of the crop and has been reported from Uttar Pradesh, Punjab and Haryana (Dhaliwal 1980). The adult can be identified by straw colour and two deep brown spots on each forewing. The pest completes its life cycle within 29 days and can have three generations in one cropping season of clusterbean (Ram and Verma 1991). The larva of the insect crawls on the plant and attached near the midrib or thick vein of the leaf by spinning a web. It can be controlled by foliar application of Carbaryl and dusting of Malathion.

Bemisia tabaci: It is a small harmful polyphagus pest recorded on very wide range of cultivated and wild plants (Basu 1995) and considered as a quarantine pest. It is a plant sap-sucking insect and has a broad host range (Othman et al. 2002). Its presence can be noticed as yellow/white larval scales. Infestation of *B. tabaci* can be seen as chlorotic spotting, vein and leaf yellowing, yellow blotching and mosaic with curling leaves. In addition to direct damage by its feeding activity, it carries various viruses and can cause considerable loss due to virus transmission. There are number of clusterbean genotypes which were screened to find out the source of resistance against *B. tabaci* (Singh et al. 1996).

The higher effectiveness was observed with the application of clothianidin and thiamethoxam against jassid and whitefly, whereas imidacloprid against jassids and spiromesifen against whitefly. Fipronil, acephate and carbosulfan effectively managed thrips on clusterbean.

Amsacta moorei: It is commonly known as red hairy caterpillar and has been reported from sandy areas of Rajasthan, Haryana, Gujarat and Punjab (Saini 1993). The pest remains active from the last week of June to almost second week of October. The larval stage of the pest is destructive to the plant that feeds rapidly and its feeding tendency increases with the age.

Manual picking and mass trapping are quite effective measures for controlling the pest (Saini and Verma 1991). Deep ploughing and flooding after harvest of the crop kill the pupae present in the soil and prevents carryover of the pest. Dosages of sevimol, carbaryl, chloropyriphos, quinalphos, monocrotophos and triazophos have been suggested for control of different instar of larvae (Saini 1995).

Silosoma obliqua: It is commonly known as Bihar hairy caterpillar and is a sporadic and polyphagous pest causing considerable damage to the crop. The pest appears in the young crop and feed leaves, soft portion of stem and branches rapidly. The adult moth measuring of 40–50 mm is characterized with pink or dull yellow head, the thorax and under side body of the moth has black dots. It remains active during the months of July to November.

To prevent the population build-up of the pest in the crop, the leaves bearing egg masses should be collected and destroyed so that larval stage could not be developed. The severe infection of all the stages can be managed by spraying endosulfan or quinalphos or dichlorvos.

5.7.2 Gall Midges

Gall midges are small, inconspicuous flies and the adult gall midges have small delicate bodies, usually only 0.5–3 mm long, rarely up to 8 mm long and long slender antennae. They resemble the fungus gnats and flies in certain other families. Adults of gall midges are not harmful as they do not feed. Adults have very short lives, sometimes only several hours to 1–4 days. Larvae are tiny yellow to orange maggots that can't be seen unless the galls are cut to open. The larvae of gall midges are probably best known for inducing abnormal plant growths called galls that cause economic damage (Patel and Chari 1970).

Asphondylia cyamopsii **n. sp.**: The adults of *A. cyamopsii* are 3–4 mm long strong fliers having dusky drab colour. The females lay eggs in the ovary of the flower buds and pods. On hatching, the maggot comes out and starts feeding on the ovules, the adult emerges out from the hole of the gall. It was reported on the garden clusterbean by Rao (1917) from India. Patel and Chari (1970) reported heavy infestation in the commercial crop of clusterbean due to gall midge induced by pod gall formation. Gagné and Wuensche (1986) reported *A. websteri* as an invasive pest of clusterbean. The maggots of the pest is responsible for production of gall which becomes bigger and hard in due course of time. Due to infestation the stigma does not develop normally and remains like a hook, which is a peculiar diagnostic character of gall midge infestation (Chari 1971).

Manual removal of infested buds and early sowing avoid the incidence of the pest. Effective control of the pest has been achieved with spray of carbaryl or phosphamidon and monocrotophos at flower initiation stage (Krishna Kumar and Sardana 1988).

Contarinia texana: It is one of the primary insect pests of clusterbean in the Southwest Oklahoma (Undersander et al. 1991) and caused up to 30 % loss in seed production due to heavy midge infestations. Rogers (1972) reported presence of this midge associated with clusterbean in Texas and Oklahoma. The infestations are generally heavier in fields with sandy or sandy loam soils. The adult female midge deposits eggs in the developing buds. The crop is damaged by the larvae, which develop in the flower buds. Infested buds eventually dry up and fall down. The developed larvae drop from the buds to ground to pupate. The insect may have several generations in one year. Field inspection should be carefully done primarily between 45 and 90 days after emergence, when the crop is producing buds.

5.7.3 Thrips

Thrips are multinational in distribution and do not cause serious damage but can be severe and sometimes have serious implications. Thrips of various species (*M. distalis*, *Ayyaria chaetophora*, *Caliothrips phaseoli*, *Taeniothrips distalis*, *Frankliniella schultzei*, etc.) have been reported on the clusterbean crop.

Megalurothrips distalis: *M. distalis* usually occurs on flowers and damages the anthers, stigma and sometimes leaves (Butani 1980). It causes curling of the pods by feeding in the flowers during pod formation, malformation of the young plant by feeding at the growing point and curling of leaves by feeding on the undersides. In good climatic conditions the plants may recover itself from damage but the late-sown crops are often severely injured and can not recover the damage. Butani (1980) mentioned that nymphs and adults infest the flowers and feed on pedicles, sepals, petals and even on the stigma of the flowers of clusterbean.

Ayyaria chaetophora: It is a polyphagous species of thrips damaging leaves of clusterbean (Ananthakrishnan 1971) and commonly known as leaf feeding thrips. The thrips causes damage by scrapping and sucking the sap from the leaf tissues and completes its life cycle in 15–20 days. The leaves attached by many thrips develop a silvery lustre due to discoloured emptied cell and fell down prematurely.

Caliothrips phaseoli: It is also a polyphagous species having similar symptoms caused by *A. chaetophora*. It has a life cycle of 11–14 days from first instar to adult having several generations in a year. Manipulation in date of sowing, date of harvesting, adequate moisture and weed management reduce the severity of the attack of thrips. Jotwani and Bhutani (1977) recommended dimethoate, phosphamidon, formothion, thiometon or endosulfan to control the pest on its appearance.

5.7.4 Leaf Hoppers

Empoasca kern: The jessid or leafhoppers are widely distributed throughout the country, appears at early stage of the crop and achieves the peak during third week of August. It is one of the destructive pests in north-western region of the country and both the nymphs and adults cause serious damage to the crop. The infested leaves turn pale and eventually dry up and fall. Spray of systemic insecticides, viz. dimethoate or formothion, phosphamidon or oxydemeton methyl is reported to be effective in controlling jassid activity. These insecticides can be repeated at intervals of 10–14 days for complete check of the pest.

Exitianus indicus: Both the nymphs and adults of the pest suck sap from leaves and tender shoots of the plant but damage is almost negligible. Adults of the pest are active, slender, having head, thorax and scutellum of greenish colour.

Holotrichia consanguinea: The pest is commonly known as white grubs, beetle grubs, chafer beetles or June beetles and is a predominant species particularly in Haryana, Rajasthan, Punjab and Himachal Pradesh (Brar and Sandhu 1981) on different host range including clusterbean.

Seed treatment with imidacloprid is used for the control of white grubs in clusterbean. Spraying of Carbaryl, Monocrotophos or quinalphos give effective control on the adult pest. Application of Chlorpyrifos or quinalphos with irrigation water gives better control over the standing crop (Gupta and Yadava 1987). Seed treatment with these insecticides also gives better results.

Yellow mite (*Polyphagotarsonemus latus* and *P. datus*): *P. latus* is a microscopic species live in colonies and infest mostly the top leaves of many plant species including clusterbean. *P. latus* infestation can cause stunting and twisting of the leaves and flowers. The mites prefer areas of high humidity and low temperature and are the most prevalent pest on the green house plants in temperate and subtropical areas (Brown and Jones 1983). It can be controlled by removing and destroying infested plants, and spraying with a miticide.

5.8 Nematodes

Root-knot nematode (*Meloidogyne incognita and M. javanica*), Reniform nematode (*Rotylenchulus reniformis*), Pigeon pea cyst nematode (*Heterodera cajani*) are the main nematodes affecting the crop (Ananymous 2002).

5.8.1 Root-Knot Nematodes

Root-knot nematodes are sedentary endo-parasites of diverse crops and are one of the most important polyphagous pests of clusterbean. Twelve species of this nematode is reported from India but five of them, viz. *M. incognita*, *M. javanica*, *M. hapla*, *M. arenaria* and *M. graminicola* are wide spread (Musarrat et al. 2007). The nematodes induce root knot or root gall of varying sizes on the root systems and the severity can be easily determined by pulling a plant or digging around the root. The above ground effects can be recognized as lack of vigour, stunted growth, yellowing of leaves and wilting under water stress conditions.

The most effective and inexpensive way of reducing destructive effects of root-knot nematodes are the crop rotation with non-host crops and use of resistant or tolerant varieties. Soil solarization can be effective for the sterilizing of soil during favourable environmental conditions.

5.8.2 Pigeon Pea Cyst Nematode

The nematode completes its life cycle in 3 weeks and reproduces rapidly. The optimum temperature for its life cycle is 25–30 °C. Clusterbean was first reported as the host of pigeon pea cyst nematode (Bhatti and Gupta 1973). The main symptoms of the infection are stunted growth and reduced yield due to damage to the root system of host plant. Flowers and pods are reduced in size and number and the root system poorly developed. It has been widespread in clusterbean growing areas of Haryana (Walia et al. 1985) and Rajasthan (Dutta et al. 1987). Elimination of cyst nematodes from infested field is very tough task. An integrated control approach involving summer ploughing, crop rotation, seed treatment, limited applications of chemicals and the use of resistant varieties may be adopted for the control of cyst nematodes.

5.9 Stem Weevils

Stem weevils (species of *Alcidodes*, *Blosyrus*, *Cyrtozemia* and *Myllocerus*), especially *A. bubo* (Fab.), *A. collaris* (Pasc.), *A. fabricii* (Fab.), *A. pictus* (Bohem), *Blosyrus inequalis* (Bohm.), *Cyrtozema cognata* (Marsh.), *C. dispar* (Marsh.) and *M. undecimpustulatus* var. *maculosus* Desbr. (*M. maculosus*) (Butani 1980) have been reported on clusterbean. Out of which *A. bubo* and *M. maculosus* are more abundant as compared to others. The pest makes numerous holes in the leaves and stem and weakens the plant. The grub of the weevils are more destructive as compared to adult and bore the stem making irregular galleries and gall like swellings at the place of injury. Besides mechanical method of control use of insecticides like Malathion or Trichlorpphon or Carbaryl help in the control of *A. bubo*. Regular inter-culture and deep ploughing of the field are suggested to control the pest.

5.10 Aphids

Acryrthosiphon pisum and *A. craccivora* are the major aphids associated with clusterbean. These aphids are the sucking pests as well as carrier of the viruses thus damaging the crop from both the ways (COPR 1981). They are polyphagous insects damage the young plants of clusterbean in early stages from July to September (Chhabra et al. 1993; Chhabra and Kooner 1989). Their adults are pear-shaped, green yellow or pink in colour with long prominent cornicles and have parthenogenetic type of reproduction. Male pests are rare with winged or without wings (Butani and Jotwani 1984).

Intercropping with non-leguminous crop, manipulation in date of sowing and spacing help in the control of the aphid (Chhabra et al. 1993; Lal et al. 1989). Aphicide, dimethoate and thiometon are reported as effective control over the aphids.

5.11 Wasp (*Eurytoma* sp.)

The larvae of the wasp form galls on the pods of the plant and do not affect the stem or leaves. Gaur and Singh (2014) reported incidence of gall inducing wasp on clusterbean crop in Southwest Haryana from last week of August to first week of October. After hatching, the larvae rapidly form gall and the pupation occurred within the gall and each gall contained a single pupa. The pupal period ranged from 8 to 10 days. Due to gall formation, pods do not open and there is no seed formation.

5.12 Integrated Disease Management

Integrated disease management (IDM) approach has been accepted globally for achieving sustainability in the production of the crops. It has various advantages including safety to environment, pesticide free commodities, low input based crop production programmes, etc. (Kumar 2014). Although clusterbean is affected by few diseases and pests, an IDM approach for the crop is summarized as under:

- Deep summer ploughing to expose resting stages of pests and soil-borne nematodes.
- Soil solarization in the month of May is found the most effective measure in checking soil-borne pathogens.
- Timely sowing of the crops to avoid peak infestation of pests.
- Growing of short duration and recommended varieties of the region to escape the disease incidence.
- Crop should be sown in rows at optimum depths under proper moisture conditions for better establishment. After good establishment, clusterbean performs better and copes with the pathogens and insects pest.
- Crop rotation with cereals helps in reducing beetle, thrips and nematode populations.
- Removal of any crop residues is essential to prevent population of soil-borne nematodes.
- Balanced doses of fertilizers should be used.
- Roguing of affected plants particularly virus-infected plant should be timely carried out.
- Field sanitation by resorting timely hand hoeing/hand weeding.
- Hand collection and destruction of egg mass, larvae and infected leaves.
- Seed treatment with hot water (56 °C) for 10 min before sowing.
- Seed treatment with *Trichoderma viride* seed to prevent the crop from soil borne fungal infection.

- Soil application of phorate/carbofuran may be followed before sowing for protection against thrips in summer crop and for protection against stem fly and jassids during rainy season crop.
- Spray of streptocycline twice at 15 days interval at the appearance of bacterial leaf blight may be done.
- Spray of Di M-45 thrice at an interval of 15 days at the appearance of Alternaria blight and also one spray of imidacloprid may be done to check insect pests.
- Kitazin seed along with Chloropyriphos and streptocyline may be used to check the presence of insect pests, fungi and bacteria during storage.
- The crop should be sprayed with wettable cresan and cupravid at an interval of 15 days or diluted Sulphur at the appearance of powdery mildew disease.
- Alachlor as pre-emergence herbicide may be used for weed management.
- The seeds may be treated with carbofuron/carbosulfan to reduce the root-knot, cyst and reniform nematode problems.
- Only required quantities of approved labelled pesticides for single application in specified areas should be purchased and it should be stored away from food, feed and fodder.

References

Akem C, Lodha S (1997) Integrated management of soil borne diseases of legumes in dry regions. In: Proceedings of Indian phytopathological society-golden jubilee, New Delhi, 10–15 Nov, pp 508–511

Ananthakrishnan TN (1971) Thrips (Thysanoptera) in agriculture, horticulture and forestry-diagnosis, bionomics and control. J Sci Ind Res 30:113–146

Ananymous (2002) Forage crops-pest spectrum and description. In: Faruqui SA, Pandey KC, Singh JB (eds) Forage plant protection Indian Grassland and Fodder Research Institute, Jhansi, India, p 14

Anil PSK, Kadian SP et al (2012) Inheritance of bacterial leaf blight resistance in Clusterbean. Forage Res 38(3):182–183

Arya HC (1956) On a new leaf spot disease of gawar (Cyamopsis tetragonoloba L. Taub.) caused by Myrothecium roridum Tode ex Fr. Indian Phytopathol 9:174–181

Arya HC (1959) On the occurrence of physiological strains in Myrothecium rosidum. Indian Phytopathol 12:164–167

Ayub M, Mirza JH, Shakir AS (1995) Physiological studies of Alternaria cyamopsidis rang and Rao causing leaf blight of guar (Cyamopsis tetragonoloba L. Taub). Pak J Phytopathol 7:41–43

Basu AN (1995) Bemisia tabaci (Gennadius) Crop pest and principal whitefly vector of plant viruses. Westview Press, Boulder, pp 173–174

Begum H, Madhusudan T (1989) Effect of mosaic virus infection on growth and yield of clusterbean. Indian J Mycol Res 27:191–194

Bhatti DS, Gupta DC (1973) Guar an additional host of Heterodera cajani. Indian J Nematol 3:160

Brar KS, Sandhu AS (1981) Note on the effect of dates of sowing on the damage of Holotrichia consanguinea Blanchard to groundnut. Indian J Agric Sci 51:466

Brown RD, Jones VP (1983) The broad mite on lemons in Southern California. California Agric 37(7/8):21–22

Butani DK (1979) Insects and fruits. Delhi, Periodical Expert Book Agency, p 415

Butani DK (1980) Insect pests of vegetables and their control-clusterbeans. Pesticides 14(10):33–35

Butani DK, Jotwani MG (1984) Insects in vegetables. Periodical Expert Book Agency, Delhi, p 356

Butler EJ (1918) Fungi and diseases in plants. Thacker Spink and Company, Calcutta

Chahal SS, Sahoo MS, Gill KS (1980) Source of resistance against bacterial blight of guar (*Cyamopsis tetragonoloba*). Indian J Mycol Plant Pathol 9:103–104

Chand JN, Verma PS (1968) A leaf spot disease of *Cyamopsis tetragonoloba* caused by *Curvularia lunata* in India. Indian Phytopathol 21:239–240

Chandra U, Saxena SK (1990) Occurrence of *Sphaerotheca fuliginea* (Schelecht ex. Fr.) Poll on guar bean (*Cyamopsis tetragonoloba* L. Taub.) in and around Aligarh. Natl Acad Sci Lett 13:261–262

Chari MS (1971) Nature and extent of damage caused by the larvae of clusterbean pod gall midge *Asphondylia cyamopsii* n sp. on different varieties of clusterbean (*Cyamopsis tetragonoloba* L. Taub.). Madras Agric J 58:892–893

Chestor KS, Cooper WE (1944) A lethal virus of guar (*Cyamopsis psoralioides* DC). Phytopathology 34:998

Chhabra KS, Kooner BS (1989) Sources pf whitefly *Bemisia tabaci* G and yellow mosaic virus resistance in mung bean *Vigna radiata*. Tropical Grain Legume Bull 19:26–29

Chhabra KS, Lal S, Kooner BS et al (1993) Insect pests of pulses-identification and control manual. PAU Ludhiana and Directorate of Pulses Research, Kanpur, p 88

COPR (1981) Pest control in tropical grain legumes. Center for Overseas Pest Control, London, pp 127–206

Dhaliwal JS (1980) Important insect pests of guar in the Punjab state. Guar Newsletter 1:18

Dutta S, Trivedi PC, Tiagi B (1987) Nematodes of guar and mung in some areas of Rajasthan, India. Int Nematol Network Newsletter 4:12–16

Fulton RW (1948) Hosts of tobacco streak virus. Phytopathology 38:421–428

Fulton RW (1950) Variants of tobacco necrosis virus in Wisconsin. Phytopathology 40:298–305

Gagné RJ, Wuensche AL (1986) Identity of *Asphondylia* (Diptera: Cecidomyiidae) on Guar, *Cyamopsis tetragonoloba* (Fabaceae) in the Southwestern United States. Ann Entomol Soc Am 79:246–250

Gandhi SK, Chand JN (1985) Yield losses in guar due to bacterial blight caused by *Xanthomonas campestris* pv Cyamopsidis. Indian Phytopathol 38:516–519

Gandhi SK, Chand JN, Saini ML (1978a) Reaction of genetic stock of guar (*Cyamopsis tetragonoloba* L. Taub.) to bacterial blight (*Xanthomonas cyamopsidis*) and its incidence in relation to date of sowing. Forage Res 4:159–162

Gandhi SK, Chand JN, Saini ML (1978b) Screening of clusterbean genotypes for resistance to leaf spot caused by *Alternaria cyamopsidis*. Forage Res 4:169–171

Gaur RK, Singh B (2014) Guar (*Cyamopsis tetragonoloba*), a new host of gall inducing wasp, *Eurytoma* sp. in South West Haryana. Ann Plant Prot Sci 22(2):437–438

Gupta V (1997) Management of Alternaria leaf spot in guar. MSc thesis, CCS HAU, Hisar, India, 60 pp

Gupta BM, Yadava CPS (1987) Chemical control of white grub, *Holotrichia consanguinea* Blanchard infesting groundnut. Indian J Agric Sci 57:742–744

Harris DC, Funnell PS, Davies MK et al (1986) Diseases of strawberries. Report of East Malling Res Station, p 121

Jotwani MG, Butani DK (1977) Insect pests of leguminous vegetables and their control. Pesticides 11(10):35–38

Kalaskar SR, Shinde AS, Dhembre VM et al (2014) The role of peroxidase and polyphenoloxidase isoenzymes in resistance to bacterial leaf blight (*Xanthomonas campestris* pv. cyamopsidis) in clusterbean (*Cyamopsis tetragonoloba* (L.) Taub.). J Cell Tissue Res 14(3):4577–4580

Karwasara SS, Chand JN (1982) Screening of genetic stock of guar (*Cyamopsis tetragonoloba*) to bacterial blight (*Xanthomonas campestris* pv. *Cyamopsidis*). Indian J Mycol Plant Pathol 12:229

Kernhamp MF, Hemerick GA (1953) The relation of Ascochyta to Alfalfa seed production in Minnesota. Phytopathology 43:378–383

Khan SA (1958) Powdery mildew of Tandojam. Pak J Sci Res 10:82

Kothari KL, Bhatnagar MK (1966) Evaluation of fungicides against crop diseases, on *Colletotrichum capsici* causing blight of guar. Indian Phytopathol 19:116–117

Krishna Kumar NK, Sardana HR (1988) Pest management in solanaceous vegetables. Inl Annual Report IIHR, Bangalore 64

Kumar D (2005) Status and direction of arid legumes research in India. Indian J Agric Sci 75(7):375–391

Kumar S (2014) Plant disease management in India: advances and challenges. Afr J Agric Res 9(15):1207–1217

Kumar S, Sharma S, Kumar N et al (2008) Management of dry root rot of clusterbean caused by *Rhizoctonia bataticola*. Forage Res 34(3):160–164

Kumar J, Kumar A, Roy JK et al (2010) Identification and molecular characterization of begomovirus and associated satellite DNA molecules infecting *Cyamopsis tetragonoloba*. Virus Gene 41:118–125

Lal SS, Yadava CP, Dias CAR (1989) Effect of planting density and chickpea cultivars on the infestation of the black aphid (*Aphis craccivora* Koch.). Madras Agric J 76:461–462

Le Beau FJ (1947) A virus induced top necrosis in bean. Phytopathology 37:434

Lodha S (1984) Varietal resistance and evaluation of seed dressers against bacterial blight of guar (*Cyamopsis tetragonoloba*). Indian Phytopathol 37:438–440

Lodha S (1996) Influence of moisture conservation techniques on *Macrophomina phaseolina* population, dry root rot and yield of clusterbean. Indian Phytopathol 49:342–349

Lodha S, Anantharam K (1993) Effectiveness of streptocycline spray schedule on the bacterial blight intensity and yield of clusterbean. Ann Arid Zone 32:237–240

Lodha SK, Mawar R (2002) Diseases. In: Kumar D, Singh NB (eds) Guar in India. Scientific Publishers (India), Jodhpur, Rajasthan, pp 127–147

Mathur SB (1962) Effect of soil moisture on root rot of guar (*Cyamopsis psoralioides* DC) and wilt of gram (*Cicer arietinum* L.) caused by *Sclerotium rolfsii* Sacc. Agra Univ J Res 11:295–302

Mathur RL, Mathur BN, Sharma BS (1972) Relative efficacy of fungicides for the control of *Alternaria cyamopsidis* causing leaf spot of guar (*Cyamopsis tetragonoloba*). Indian J Mycol Plant Pathol 2:80–81

Mishra JN (1963) Ozonium wilt of guar and cucurbit plants in Bihar. Nature 172:209–210

Musarrat AR, Zarina B, Shahina F (2007) New host of root-knot nematode in Pakistan. Pak J Nematol 25:341

Orellana RB (1967a) Reaction of guar to strains of Tobacco ring spot virus. Phytopathology 57:791–792

Orellana RG (1967b) Leaf spot of guar caused by *Pseudomonas syringae* in the United States. Plant Dis Rep 51:182–184

Orellana RG, Thomas CA, Kinman ML (1965) A bacterial blight of guar in United States. FAO Plant Prot Bull 13:9–13

Othman AKM, Shuaib OS, Sattar MHA (2002) Survey for host plants of whitefly, *Bemisia tabaci* (Gnn.) in Abyan and Tuban Delta at southern coastal plain. Univ Aden J Nat Appl Sci 6(3):497–504

Pareek V, Varma R (2014) Phytopathological effects and disease transmission in clusterbean seeds grown in Rajasthan. Indian J Plant Sci 3(2):26–30

Patel KH, Chari MS (1970) New record of gall midge Asphondylia sp. as a serious pest of clusterbean (*Cyamopsis tetragonoloba* L. Taub.). Indian J Entomol 32:90–91

Patel MK, Dhande GW, Kulkarni YS (1953) Bacterial leaf spot of (*Cyamopsis tetragonoloba* L. Taub.). Curr Sci 22:183

Patel PS, Patel IS, Panickar B (2011) A new record of whitefly, *Acaudaleyrodes rachipora* attacking clusterbean in India. Insect Environ 17(3):136–137

Prasad N (1945) Studies on the root rot of cotton in Sind II. Relation of root rot of cotton with root rot of other crops. Indian J Agric Sci 14:388–391

Prasad N, Desai MV (1952) Fusarium blight of clusterbean. Curr Sci 21:17–18

Purkayastha S, Kaur B, Dilbaghi N et al (2006) Characterization of *Macrophomina phaseolina*, the charcoal rot pathogen of clusterbean, using conventional techniques and PCR-based molecular markers. Plant Pathol 55:106–116

Ram P, Verma AN (1991) Biology of *Dichomeris ianthus* Meyr-a pest of clusterbean (*Cyamopsis tetragonoloba* L. Taub.) in Haryana. Agric Sci Digest 11(4):181–184

Rangaswami G, Gowda SS (1963) On some bacterial diseases of ornamental and vegetables in Madras state. Indian Phytopathol 16:74–85

Rangaswami G, Rao V (1957) *Alternaria blight* of clusterbean. Indian Phytopathol 10:18–25

Rao R (1917) Clusterbean (*Cyamopsis psoralioides*) (Guar-Hindi). In: Proceeding of II entomology meeting Pusa Calcutta, India, pp 60–61

Ratnam CV, Pandit SV, Rao KC (1985) Evaluation of some systemic and non-systemic fungicides against powdery mildew of clusterbean. Pesticides 19:36–37

Raychaudhuri SP, Nariaini TR, Das CR (1962) *Cyamopsis tetragonoloba* L. Taub.-a local lesion host of sunhemp mosaic virus. Indian Phytopathol 15:79–80

Reddy MS, Rao AA (1971) Benlate: a systemic fungicide highly effective against powdery mildews. Indian Phytopathol 24:196–197

Ren YZ, Yue YL, Jin GX et al (2014) First report of bacterial blight of guar caused by *Xanthomonas axonopodis* pv.*cyamopsidis* in China. Plant Dis 98(6):840

Rogers CE (1972) Midge associated with Guar in Texas and Oklahoma. Ann Rnyomol Dov Smrt 65:1203–1208

Saharan MS, Saharan GS (2004) Influence of weather factors on the incidence of *Alternaria blight* of clusterbean (*Cyamopsis tetragonoloba* (L.) Taub.) on varieties with different susceptibilities. Crop Prot 23:1223–1227

Saharan MS, Saharan GS, Singh S (2001) Effect of *Alternaria blight* severity on fodder quality of clusterbean. Forage Res 27(3):221–224

Saini RK (1993) Development and survival of red hairy caterpillar, *Amsacta moorei* Butler on some cultivated plants and weeds in Haryana. J Insect Sci 6:64–68

Saini RK (1995) Evaluation of some promising insecticides against red hairy caterpillar, *Amsacta moorei* Butler. J Insect Sci 8:220–222

Saini RK, Verma AN (1991) Effectiveness of light traps in suppressing population of red hairy caterpillar, *Amsacta moorei* Butler. Haryana Agric J Res 21:250–252

Schut M, Rodenburg J, Klerkx L et al (2014) Systems approaches to innovation in crop protection. A systematic literature review. Crop Prot 56:98–108

Sekar R, Sulochana CB (1986) Studies on the cross protection behaviour on TMV strains in a local lesion host. Indian J Plant Pathol 4:33–39

Sharma SR (1983) Effect of fungicides on the development of *Alternaria blight* and yield of clusterbean. Indian J Agric Sci 53:932–935

Sharma SR (1984) Effect of different cropping seasons and tridemorph sprays on powdery mildew and yield of clusterbean. Z Pflanzenphysiol 2:81–89

Sharma S, Pathak DV, Kumar S et al (2005) Efficacy of plant formulations and bioagents in controlling root rot of clusterbean. Forage Res 31(2):103–105

Shivanna MB, Shetty HS (1986) Myrothecium pod spot of clusterbean and its significance. Curr Sci 55:574–576

Singh RS (1951) Root rot of guar. Sci Cult 17:131–134

Singh RS (1954) Effect of mixed cropping on the incidence of root rot, wilt and associated disease of *Cyamopsis psoraloides*. Agra Univ J Res 3:359–373

Singh I, Chouhan JS (1973) Seed mycoflora of guar (*Cyamopsis tetragonoloba*) and their effect on germination and growth of seedlings. Indian J Mycol Plant Pathol 3:86–92

Singh SD, Prasada R (1972) Studies on physiology and control of *Alternaria cyamopsidis* the incitant of blight disease of guar. Indian J Mycol Plant Pathol 3:33–39

Singh B, Singh RS (1957) Incidence of seedling rot, root-rot and wilt of *Cyamopsis psoraloides* under manured and unmanured conditions. Agra Univ J Res 6:7–14

Singh J, Swarup J (1986) Control of bacterial blight of clusterbean by seed treatment and foliar spray. Pesticides 20:23–24

Singh J, Swarup J (1987) Epidemiology of bacterial blight of clusterbean caused by *Xanthomonas campestris* pv. *cyamopsidis*. Farm Sci J 2:91–92

Singh G, Gupta RBL, Dela GG (1974) Efficacy of fungicides and varietal resistance of clusterbean (*Cyamopsis tetragonoloba*) against leaf spot disease caused by *Curvularia lunata*. Indian Phytopathol 27:234–236

Singh SP, Singh JV, Singh VP (1996) Screening of clusterbean genotypes for resistance of whitefly, *Bemisia tabaci* Genn. Forage Res 22:59–62

Smith OF (1945) Parasitism of *R. solani* from alfalfa. Phytopathol 35:832–834

Solanki JS, Singh RR (1976) Note on the fungicidal control of powdery mildew of clusterbean. Indian J Agric Sci 46:241–243

Sowell G (1965) The effect of seed treatment on seed borne pathogen of guar. Plant Dis Rep 49:895–897

Srivastava DN, Rao YP (1963) Bacterial blight of guar. Indian Phytopathol 16:69–73

Streets RB (1948) Disease of guar (*Cyamopsis psoraloides*). Phytopathology 38:918

Suryanarayana D, Bhombe BD (1961) Studies on fungal flora of some vegetable seeds. Indian Phytopathol 14:30–41

Undersander DJ, Putnam DH, Kaminski AR et al (1991) Guar. In: Alternative field crops manual, University of Wisconsin-Madison. www.hort.purdue.edu/newcrop/afem/guar.html

Vani V, Madhusudan T, Rao RDVJP (1986) A new mosaic disease of clusterbean. Indian J Virol 2:222–230

Vasudeva RS (1960) Report of division of mycology and plant pathology. Science report, Agricultural Research Institute, New Delhi 1957–1958, pp 111–130

Verma GS, Verma JP, Saxena PN (1962) Top necrosis of *Cyamopsis tetragonoloba* (L.) Taub. Proc Indian Acad Sci 32:287–292

Vijayanand GK, Shylaja MD, Krishnappa M et al (1999) An approach to obtain specific polyclonal antisera to *Xanthomonas campestris* pv. *cyamopsidis* and its potential application in indexing of infected seeds of guar. J Appl Microbiol 87:711–717

Wadhwa N, Joshi UN, Mehta N (2014) Zinc induced enzymatic defense mechanisms in rhizoctonia root rot infected clusterbean seedlings. J Bot doi.org/10.1155/2014/735760

Walia RK, Bajaj HK, Gupta DC (1985) Prevalence of pigeon pea cyst nematode, *Heterodera cajani* Koshy, in Bhiwani district (Haryana) along with a new host record. Haryana Agric Univ J Res 15:463–464

Wojtaszek P (1997) Oxidative burst: and early plant response to pathogen infection. Biochem J 322:681–692

Yogendra S, Kushwaha KPS, Chauhan SS (1995) Epidemiology of Alternaria leaf blight of clusterbean caused by *Alternaria cyamopsidis*. Ann Plant Prot Sci 3:171–172

Chapter 6
Physiology and Abiotic Stresses

Abstract Clusterbean is a robust crop and copes with almost all the physiological and abiotic stresses; however, for good crop productivity these stresses should be properly addressed. The physiological aspects associated with seed maturation, seed coat colour, germination, seedling/plant growth, seed yield and seed storage in clusterbean are taken in hand.

6.1 Introduction

Plant growth is greatly affected by environmental stresses, such as drought, high salinity and low temperature and pathogen infection during the life cycle. To encounter these stresses, plants have evolved mechanisms to increase their tolerance through physical, molecular and cellular adaptations. Abiotic stresses have become an integral part of crop production under changing climate scenario (Mittler and Blumwald 2010). Among abiotic stresses drought, water stress and salinity are the main abiotic factors for yield reduction (Munns and Tester 2008; Reynolds and Tuberosa 2008).

6.2 Seed Maturation

Seed maturation refers to the morphological, physiological and functional changes that occur from the time of anthesis to the harvest of seeds (Khattra and Singh 1995). Maturity is critical and the most important factor that determines the size, quality, planting value and storability of the seed (Jerlin et al. 2001). Physiological maturation occurs commonly in seeds to gain the reproducing capacity of the younger generation and this normally coincides with the attainment of maximum dry weight when the flow of nutrients to the seed from the mother plant is ceased (Harrington 1973). Khattra and Singh (1995) observed that accumulation of dry matter with loss of moisture is one of the characteristic features that could be

© Springer Science+Business Media Singapore 2015
R. Pathak, *Clusterbean: Physiology, Genetics and Cultivation*,
DOI 10.1007/978-981-287-907-3_6

expressed during seed development and maturation in any crop. With advancement in maturation, the decrease in moisture content of pod at a faster rate could be due to the replacement of osmotic materials with low hydration capacity (Harrington 1973). Renugadevi et al. (2006) noticed reduction in dry weight of clusterbean pod and seed on maturation. The maturation process proceeds with water loss at various degrees depending on the atmospheric condition. The change in pod and seed colour is considered as a promising visual index of seed maturation (Carlson 1973). Based on physical, physiological, biochemical and visual indices, cluster-bean seeds reach their physiological maturity at 18.2 % of seed moisture content (Renugadevi et al. 2006).

6.3 Seed Coat Colour

The seeds of clusterbean with black seed coat are believed to be of low quality (Bhatia et al. 1979), but these seeds may be recommended for planting or gum extraction with little loss in stand or gum yield. Liu et al. (2007) examined the effect of seed coat colour on water uptake, germination and quantity of seed com-ponent of clusterbean and found that the rate of water uptake for black coloured seeds was greater than that of dull white coloured seeds; besides these black seeds had higher germination compared to dull white coloured seeds. Seed coat of clus-terbean is affected by environment and black seeds are reported to have gallic acids, 2- ,3- ,4-trihydroxybenxoic acids, ferric ions, and galactose (Bhatia et al. 1979) and has higher incidence of *Aspergillus flavus* content compared to dull white coloured seeds (Paredes-Lopez et al. 1989). Dark coloured seeds appeared to be scarified by nature while light coloured seeds retained the hard seed coat and the seed viability may vary with seed colour or age (Musil 1946). Lohan and Jain (2007) sorted out the off-coloured seed by digital colour sorter for improv-ing the colour purity and seed quality and found them to be of poor quality. Seth and Padmavathi (2003) studied the seed quality along with the expression of seed vigour and hard seededness in relation to seed coat colour in fodder clusterbean and reported that the seed quality in terms of germination parameters was better in whitish grey seed coat colour category, while maximum number of hard seeds was observed in greyish black category.

6.4 Germination, Seedling Growth and Storage

Seed quality of clusterbean is affected by improper storage conditions, such as presence of high temperature and relative humidity that may lead to complete loss of viability (Doijode 1989a). Clusterbean seeds survive for a short period at ambi-ent temperatures and the storage of seeds for 6 months significantly reduces the germination percentage (Kalavathi and Ramamoorthy 1992). Seeds stored in kraft paper bags maintain high viability, while seed stored in glass, polyethylene bags or laminated aluminum foil pouches retain viability relatively for a longer period under ambient conditions (Doijode 1989b). When the harvest was delayed beyond the optimum seed moisture content, substantial decline in viability along with increase in seedling abnormalities was recorded (Ellis et al. 1987). On seed ageing, the germination percentage decreases and the leakage of electrolytes, soluble sugars and free amino acids increased on seed imbibitions (Doijode 1989a).

The germination in clusterbean is epigeal and completes within a week. Kalavathy and Vanangamudi (1990) reported that large seeds had higher germination and seedling vigor than smaller seeds. Clusterbean is not able to germinate at very low soil water potential, high salinity and submerged conditions (Datta and Dayal 1988). Seed germination is rapid and higher at a constant temperature of 30 °C and at alternate temperatures of 20, 30 °C for 16, 8 h (Doijode 1989a). A constant temperature of 30 °C, seed treatment with sulphuric acid (Musil 1946), sufficient imbibition, sowing depth of 15–30 cm (Zheng et al. 1980), pre-soaking in NaCl or solution of trace elements (Singh et al. 1976) are some of the practical means for obtaining prompt and uniform germination.

Kumar et al. (2014) assessed the effects of different priming treatment (hydro-, halo-, osmo- and biopriming) on physiological and field parameters of clusterbean and reported that the germination of seed lots may be enhanced by water soaking, soaking in 0.5 % KNO_3 and treatment of seeds with *Pseudomonas* culture. Similarly, treatment of seeds with 50 ppm GA_3, soaking of seeds in PEG-6000 may improve the germination percentage. Stafford and McMichael (1990) evaluated primary and lateral root length of clusterbean seedlings under several temperature and pre-germination seed treatment regime to determine optimum conditions for root development. Hot-water treated seed grown in distilled water produced maximum number of lateral roots and total lateral root length. Like many legumes clusterbean seeds are also not able to germinate at very low and very high water potential; similarly, high salinity and submergence conditions cause reduction in germination (Datta and Dayal 1988).

6.5 Factors Limiting Plant Growth

Plants require water, light, carbon and mineral nutrients for growth but abiotic stress reduces their growth and yield below optimum levels and the plant suffers from different reversible and irreversible stresses in one or another way (Skirycz and Inze 2010) depending on the tissue or organ affected by the stress. Clusterbean does not require many minerals for plant growth. In case of nitrogen, the basal dressing imparts vigour to the young seedlings and together with phosphorus, it provides a balanced nutrition to the crop growth compared to unfertilized plant (Madalgeri and Rao 1989). Application of 1 % urea increased the leaf nitrogen content, favoured better photosynthesis and resulted in higher sink size (Bahr 2007). Kumawat et al. (2015) studied the effect of foliar application of urea on the leaf pigmentation and yield of clusterbean under rainfed conditions and observed enhanced grain yield.

Nitrate application influences nitrate reductase activity in the crop and foliar spray of nitrogen increases seed germination and seedling vigour index in clusterbean (Sathiyamoorthy and Vivekanandan 1995). Phosphate application to clusterbean enhances the nitrogen percentage in nodules and their dry weight (Kathju et al. 1987). Application of zinc gives better results in plant growth and seed yield under dryland conditions. Clusterbean seeds were grown in the presence of different concentration of zinc sulphate to assess the effect of metal on germination, growth and biochemical changes and the results indicated low level of zinc concentration (10 and 25 mg/l) which showed a significant increase in the germination, seedling growth and biochemical content; whereas the higher concentrations (50–200 mg/l) decreased these parameters (Manivasagaperumal et al. 2011).

Micronutrients (zinc, boron, copper, molybdenum, etc.) are applied in little quantity, directly or through organic manures to improve the status of micronutrient in the soil for better nodulation, better growth and seed yield. Boron deficient plant of clusterbean has thin stem, poorly developed roots and reduced basal leaves, whereas copper deficient plants have young and curled leaflets along with chlorotic margins. 0.1–0.2 ppm copper, 0.1–03 ppm boron, 5–10 ppm magnesium have been found as optimum level for the growth of clusterbean (Neelkantan and Mehta 1961). A significant increase in nitrogenase activity, seed yield and nitrogen content of clusterbean was recorded with zinc (Nandwal et al. 1990), calcium and pre-soaking of seed in tryptophane followed by molybdenum spray (Bell et al. 1989).

Seed treatment has different response on clusterbean depending on its duration. Seed treatment with Dianthane-M-45 for 12 h increased plant height and yield, while 24 h treatment had adverse effect (Ramulu and Rao 1992). Sharma and Lashkari (2009) studied the response of gibberellic acid (GA$_3$), napthalene acetic acid and cycocel on growth and yield of clusterbean and found maximum plant height with 150 ppm GA$_3$, maximum number of leaves and number of branches/plant with 2000 ppm cycocel and maximum pod weight/plant and highest yield of pods with 1000 ppm cycocel. The foliar spray of ethephon increased the growth of clusterbean seedling and crude protein content (Verma and Sankhla 1976). Increased seed yield, seed weight and harvest index have been reported with 2500 ppm of B-9 growth hormone due to greater accumulation of crude protein, soluble carbohydrates, but gum content decreased with increasing concentration of the hormone (Uprety and Yadav 1985). IAA and GA$_3$ improved adverse effect of NaCl salinity on germination, root length, shoot length, fresh and dry weight of seedlings (Rathore et al. 2009).

6.6 Nitrogen Fixing Efficiency

The symbiotic nitrogen fixation by leguminous crops has an important role in sustaining crop productivity and maintaining fertility even in marginal lands in smallholder systems of arid and semi-arid areas (Serraj and Adu-Gyamfi 2004). It is sensitive to various environmental stresses, viz. drought, soil pH, temperature, waterlogging, low phosphorus and other nutrient limitations (Zahran 1999) and nodulation is influenced by various factors, viz. salt and water stress, temperatures, soil type, pH, organic matter content, rhizobial populations, nature of the host, etc., and varies from 0 to 100,000/g of soil (Rao and Venkateswarlu 1983). It is assumed that clusterbean fixes about 30–70 kg nitrogen/hectare and leaves a residual effect equivalent to about 15–20 kg nitrogen/ha (Rao 1995).

Clusterbean forms larger and irregular root nodules with certain nitrogen fixing bacteria. Nitrogen fixation process in clusterbean is highly sensitive to soil moisture stress but at the same time it has a higher capacity to recover under favourable moisture conditions. The droughted plants of clusterbean had maximum and rapid recovery of nitrogenase activity upon re-watering compared to other legumes. Besides this, comparatively higher percentage of nitrogenase activity was retained in clusterbean at all levels of stresses. The rhizobial strains associated with clusterbean nodules are slow growing, produce profuse polysaccharides and have lower activities of uricase and allatoinase (Sheoran et al. 1982). Mishra et al. (2013) isolated 15 rhizobial bacteria nodulating clusterbean from arid and semi-arid regions of Rajasthan and observed that the rhizobial isolates from hot climatic areas had greater tolerance to abiotic stress. Phosphorus application influences the process of nodulation, symbiotic nitrogen fixation, seed yield and quality of pods in clusterbean (Kathju et al. 1987). Potassium has an important role in physiological processes and yields determining process, viz. drought, pest, and disease tolerance and minimizes the water loss through transpiration (Burman et al. 2009).

Higher salinity or sodicity and NaHCO$_3$ were found as the most detrimental for nodule number and nodule weight (Garg et al. 1986). However, clusterbean rhizobium is stable at 3 % salt concentration (Yadav and Vyas 1971). The nodulation pattern and the nitrogen fixing ability do not show any relation to the total nitrogen and its consequent contribution to the grain yield (Rao et al. 1984). The un-branched varieties were reported with higher nitrogen fixation compared to branching types (Oke 1967). The nodulation by different *Rhizobium/Bradyrhizobium* strains in clusterbean is usually poor (Rao et al. 1984) and ranged between 0.93 and 5.73 in different cultivars (Khurana et al. 1978). So, efficient *Rhizobium/Bradyrhizobium* cultures should be isolated and introduced in clusterbean growing areas to improve its nodulation status, seed quality and crop productivity (Khandelwal and Sindhu 2013). Inoculation with various *Bradyrhizobium* spp. increased dry matter production and nitrogen fixation (Asher et al. 1990), the number of nodules, nodule fresh weight, plant dry weight, nitrogen fixation, total nitrogen content, seed yield and seed quality of clusterbean (Weaver et al. 1990; Elsheikh and Ibrahim 1999). Arbuscular mycorrhizal (AM) fungi are associated with clusterbean (Bala et al. 1989) and enhance the nitrogenase activity and nodulation in clusterbean crop. Arid and semi-arid soils may contain rhizobial strains well adapted to extreme environmental stresses such as drought, high soil temperature or pH (Mpepereki et al. 1997). These strains should be isolated and may be evaluated for use as clusterbean inoculants because rhizobium inoculated plants have more number of nodules, improved height, seed yield and dry matter production compared to un-nodulated plants (Stafford and Lewis 1980).

6.7 Water Stress and Drought Tolerance

Crop productivity in arid and semi-arid regions, is often restricted due to several abiotic stresses but the amount of soil moisture available to plants during the growing season is a major limiting factor for crop yield (Boyer 1982). The inadequate and erratic precipitation compels water stress in these regions. Clusterbean is a drought avoider crop (Kumar 2005) and has higher capability to recover from water stress and provides reasonable seed yield and dry matter once the stress is relieved (Garg et al. 1998). Sheoran et al. (1980) found that cumulative water stress delayed and reduced germination, decreased moisture uptake and adversely affected the hypocotyl and radical growth of clusterbean and resulted in accumulation of starch in the seedlings. The water stress imposed at flower initiation stage decreased relative water content, photosynthesis, starch and carbohydrate accumulations (Kuhad and Sheoran 1986) and is most disadvantageous and critical to the growth of plant (Vyas et al. 2001). Increasing intensities of water stress at the pre-flowering stage result in progressive and significant decline in the plant water potential, relative water content, total chlorophyll and soluble protein and nitrate reductase activities. While free proline and free amino acids showed an increasing trend with increasing water stresses. Even mild water stress at the pre-flowering stage may reduce plant growth and yield significantly (Vyas et al. 1985). Whereas water stress imposed at vegetative and pod formation stages had negative effect on growth, photosynthesis, activity of various enzymes and seed yield of the plant (Vyas et al. 2001). It was observed that early maturing or early flowering varieties/genotypes of clusterbean perform better compared to late maturing or late flowering varieties/genotypes under low rainfall conditions. The early maturing genotypes of clusterbean maintain higher plant water status and experience less metabolic changes compared to late maturing genotypes (Garg et al. 2003).

Water stress is known to induce structural and physiological alterations in the root nodules and affect their nitrogen fixing ability (Pankhrust and Sprent 1975). It significantly decreases shoot water potential, relative water content of leaves, net photosynthetic rate, total chlorophyll, starch and soluble proteins as well as nitrate reductase activity at various growth stages. Venkateswarlu et al. (1983) studied the effect of water stress on the nodulation and nitrogenase activities of clusterbean and found that increasing water stress significantly reduced the fresh weight of nodules but the number of nodules remained unchanged, while nitrogenase activities were adversely affected. Restoration of turgidity on re-watering increased the nodule fresh weight and recovery of nitrogenase activity of droughted plants, but the original weight could not be regained due to the irreversible damage of some nodules. Rapid recovery in the nitrogenase activities has been seen upon irrigation, implying the adaptability of the crop to arid regions. The activity of nitrogen fixing bacteria in legumes reduces under moisture stress situations and the root nodules degrade after the peak flowering stage (Venkatesh and Basu 2011). Bibi et al. (2014) assessed genetic associations of various seedling traits in clusterbean genotypes under water stress conditions and reported a significant correlation of shoot length, chlorophyll a/b, fresh and dry shoot weight with irrigation suggesting that the selection of drought resistance genotypes may be helpful to improve yield under water stress conditions.

Water stress induced decline in plant water potential and leaf relative water content led to reduction in total chlorophyll, starch and soluble protein contents besides accumulation of free proline to various extents depending upon the genotype and growth stage at which drought was experienced (Shubhra et al. 2004). Under water stress, leaf water potential, osmotic potential, chlorophyll and gum contents decreased and soluble sugar contents increased in clusterbean. Shubhra (2005) studied the impact of phosphorus application on leaf characteristics, nodule growth and plant nitrogen content in clusterbean and reported that phosphorus treatment may be effective to some extent for alleviating the effect of water deficit. The water stress reduces the availability and absorption of phosphorus and the phosphorus deficiency declines the leaf growth, photosynthetic rate (Brooks 1986) and also reduces the uptake rate of nitrate and its assimilation by the nitrate reductase (Pilbeam et al. 1993). Phosphorus nutrition had a significant and favorable effect on plant growth and seed yield under increasing intensities of water stress. Burman et al. (2009) explored the interactive effects of phosphorus nutrition and water stress intensities on water relations, photosynthesis, nitrogen metabolism and yield of clusterbean and suggested that beneficial effects of phosphorus application may be achieved only up to moderate levels of water stress in clusterbean. Number of reports indicates that phosphorus nutrition under water deficits increased drought resistance and improved growth and yield of crop (Gutierrez-Boem and Thomas 1999). Similar positive effects of phosphorus application on grain yield and quality compared to control plants have been reported in clusterbean (Bhadoria et al. 1997).

Under water deficit conditions, application of potassium nutrition increases crop tolerance by utilizing the soil moisture, more efficiently and has substantial effect on enzyme activation, protein synthesis, photosynthesis, stomatal movement and the maintenance of the osmotic potential and turgor regulation of the cells (Lindhauer 1987) and influences the physiological processes and yield of the crop (Lahiri and Kachar 1985). The positive effects of potassium on water stress tolerance may be through promotion of root growth accompanied by a greater uptake of nutrients and water by plants (Rama Rao 1986), reduction of transpirational water loss and through the regulation of the stomatal functioning under water stress conditions (Kant and Kafkafi 2002), which is reflected as enhanced photosynthetic rate, plant growth and yield under stress conditions (Umar and Moinuddin 2002) and maintenance of a high pH in stroma and prevention of the photo-oxidative damage to chloroplasts (Cakmak 1997). Maintenance of adequate potassium in soil improved plant water relation, photosynthesis and yield of clusterbean under simulated water stress at different developmental stage (Garg et al. 2005).

The stress tolerance of the plants can be improved with the exogenous use of stress alleviating chemicals (Farooq et al. 2009). Among the stress alleviating compounds, thiourea is an important molecule. It is highly water soluble and easily absorbed in the living tissues. Under water stress conditions or high temperature, external use of thiourea can increase potassium ion uptake (Aldasoro et al. 1981) and simultaneously improves the crop tolerance. Foliar application of thiourea improved net photosynthesis and chlorophyll content in drought stressed clusterbean crop (Garg et al. 2006). The synergistic effects of phosphorus and thiourea enhance the net photosynthesis, leaf area, chlorophyll content and nitrogen metabolism leading to significant improvement in plant growth and seed yield under water stress condition (Burman et al. 2004). Seed soaking in 0.05 % thiourea solution for 4 h. along with 0.1 % spray of thiourea at 25 and 45 days after sowing was found to ameliorate the adverse effect of drought and improve the yield in clusterbean (Yadav et al. 2004). Kumar and Kaushik (2014) studied the response of thiourea on the growth parameters of clusterbean and reported that the seed treatment as well as foliar application of thiourea brought significant improvement in growth parameters. Similar results have been reported by Meena et al. (2014) with the application of seed + foliar application of thiourea.

Use of soil microorganisms has generated great interests by reducing the crop loss from drought stress and provides sustainable solutions for crop production under drought situations. Kadian et al. (2014) studied the interactive effect of AM fungi and potassium on growth and yield under water stress and reported that AM colonization and potassium fertilizer can mitigate the deleterious effect of water stress on growth and yield in clusterbean.

6.8 Salt Tolerance

Salt is a major constraint to the global crop production, limiting the capacity of agricultural land to sustain the increasing human population and estimated that 20 % of all the cultivated land and nearly half of irrigated land is salt-affected, resulting great loss to the yield potential of crops (Flowers 2004). It reduces nutrient uptake and increases osmotic stress of plants (Abdel-Ghani 2009). Salinity of soils is increasing rapidly all over the world and is becoming more acute problem due to irrigation of crops with low quality water, leading to net accumulation of ions in the root zone (Murillo-Amador et al. 2002; Flowers 2004).

The high ionic concentration in saline soil lowers the osmotic potential and hinders the normal development of plants. The adverse effects of salinity are attributed to the osmotic effect, ion effect, alterations in ionic composition and diversion of photosynthates and nitrogenous metabolites from growth to energy supply on ion transport, osmotic adjustment, damage repair etc. (Grenway and Muns 1980). NaCl stress significantly reduces total dry matter yield and the degree of reduction in dry matter yield increased with increasing salt stress level (Ashraf et al. 2002). This reduction may be due to decreased leaf area expansion, smaller amounts of intercepted radiation, hence decreased photosynthesis (Cachorro et al. 1994). The seed yield of clusterbean is significantly related with the leaf area indicating the higher assimilation rate of photosynthesis and their translocation. HG-884, RGC-1066, RGR-07, RGC-1031, CAZG-11-1 and RGC-936 reflected less than 10 % variation in leaf water content and may be selected as promising and more resilient genotypes (Anonymous 2014). Reports suggest that seed yield and yield attributing traits are significantly reduced at salinity levels above 8 dS m^{-1} (Ashraf et al. 2002) and decreased germination percentage, delayed the germination, root length, shoot length, fresh and dry weight of seedlings (Rathore et al. 2009). Although, salinity has depressive effect on the seed germination and early seedling growth of clusterbean but the crop was found tolerant to salt stress at germination stage (Mehta and Desai 1958; Datta and Dayal 1988).

Identification of plant characteristics that enable the plant to tolerate saline soil during germination and stand establishment would be very useful for clusterbean improvement. The reports indicate that clusterbean is a moderately salt tolerant crop and a significant reduction in growth, seed yield and the galactomannan content occurs only after a threshold level of salinity (Garg and Burman 2001). Garg et al. (1996) found that NaHCO$_3$ is the most unfavourable salt accountable for shoot dry weight, pod and seed yield compared to Na$_2$SO$_4$ and NaCl, while higher salinity restricts the N, P and K uptake at different developmental stage of clusterbean (Garg et al. 1997a). Lahiri et al. (1987) observed reduced uptake of N, P and K with increasing soil salinity at different developmental stages but with the maturity of plant, the uptake was increased. Increasing NaCl concentration has detrimental effect on soluble protein, free amino acids and various enzymes of nitrogen metabolism. However, these adverse effects can be improved with the application of CaCl$_2$ (Garg et al. 1997b). Irrigation with saline water up to 80 mel^{-1} salt concentrations at flowering stage did not affect the final plant height, dry matter and seed yield of clusterbean but its higher (120 mel^{-1}) concentrations reduced these parameters (Lahiri et al. 1987). Salinity has also been reported to reduce the galactomannan content in the endosperm of clusterbean seed (Kumar and Sharma 1989).

The mineral composition of clusterbean under salt stress has also been studied. Elevated calcium levels could protect the tissues from NaCl toxicity at the sub cellular level besides improving nutrient uptake under saline conditions (Garg et al. 1997b). Supplemental calcium enhanced Ca and K contents and reduced the concentration of Na thereby increasing the K: Na ratio at all the levels of NaCl. Presoaking seed treatment (Singh et al. 1976), use of FYM and gypsum (Manchanda et al. 1988) and lifesaving irrigation under terminal drought conditions (Garg et al. 1986) are known as better management practices for higher clusterbean production under saline conditions.

The development and growth of plants under salt stress is often associated with many metabolic disorders. Salt induced metabolic hindrances have been reported by several workers in clusterbean. The adverse effects of soil salinity on the total chlorophyll, starch, reducing sugars, soluble protein, free amino acids and free proline in leaves were recorded at 4 dS m^{-1} onwards (Lahiri et al. 1987). Different sodium salts have variable effects on the leaf metabolism of clusterbean. Na$_2$SO$_4$ increased chlorophyll content while at the similar salinity level NaHCO$_3$ and NaCl decreased the chlorophyll content in the leaves (Garg et al. 1996). The salt tolerant genotypes of clusterbean did not experience more reduction in seed yield and dry matter at 10 dS m^{-1} level of soil salinity while sensitive genotypes suffers more decline at equivalent salinity (Lahiri et al. 1987). Decreased plant water potential, relative water content and increased leaf diffusive resistance of clusterbean plant has been recorded with the irrigation having increased water salinity (Garg et al. 1986).

Jat et al. (2012) studied the harmful effect of salinity and their amelioration by the application of brassinolide on physiological, biochemical traits, growth and yield of salinity tolerant and salinity susceptible cultivars of clusterbean and found that the use of brassinolide significantly increased the photosynthesis rate, transpiration rate, stomatal conductance, relative water content, chlorophyll stability index, number of pods/plant, number of seeds/pod, plant height, pod length, test weight and grain yield.

6.9 Other Abiotic Stresses

Increasing use of chemical fertilizers, toxic industrial effluents and various other human activities have badly affected the cultivated land and simultaneously the crops. The waste disposal, either from water or air pollutants is one of the major environmental problems for crop cultivation (Lagerwerff and Specht 1970) because these pollutants have an elevated concentration of heavy metals. The primary toxicity mechanisms of different heavy metal ions may be different due to their chemical properties and capacity to form organic complexes but the excess and stress of these metal ions or chelates of soluble metal may induce a series of biochemical and physiological alterations in plants resulting into membrane damage, alteration of enzyme activities and the inhibition of root growth (Lepp 1981). The early events of metal stress lead to secondary effects, viz. disturbance of hormonal balance (Bhattacharjee and Mukherjee 1994), nutrients deficiency (Adriano 1986), inhibition of photosynthesis (Clijsters and Van Assche 1985), alteration of water relation (Sheoran et al. 1991) etc. The diluted effluents of fertilizer factory having very low pH coupled with fluride, sulphate and solid concentration increased chlorophyll a, b and total chlorophyll, Mg, P and K, while undiluted effluent lowered chlorophyll contents and increased Ca, F and Na in the leaves of clusterbean (Goswami and Naik 1992). The lower concentration of paper mill effluent was in favour of germination and seedling growth, while gradual decrease in germination and seedling growth was recorded on increasing concentration of the effluent (Sharma et al. 2014). Rajan et al. (2014) studied the impact of zinc electroplating industry effluent on growth, biochemical characteristics and yield of clusterbean and reported that the values for germination and biochemical characteristics decreased with increased level of effluents. The shoot and root length of clusterbean increased at lower quantity of Zn electroplating industry effluent while the reduction in shoot and root length and total fresh and dry weight were observed after 60 days at high quantities of effluent residue. Similarly, as the level of residue increased the germination percentage decreased (Sivakumar and Rajan 2014).

Changes in different photosynthetic pigments by various metal treatments, viz. cadmium (Stobart et al. 1985), mercury (Prasad and Prasad 1987) and cadmium, zinc and mercury (Schlegel et al. 1987) have also been recorded and excessive amounts of a range of heavy metals such as copper (Mocquot et al. 1996), cobalt (Terry 1981) and zinc (Brown et al. 1972) induced chlorosis in plants which were usually similar to the symptoms of chlorosis caused due to iron deficiency. Higher amino acid and protein content was recorded due to heavy metals like copper, zinc, mercury, lead and cadmium (Nag et al. 1981), lead, cadmium, copper and zinc (Kastori et al. 1992) and cadmium and lead (Bhattacharjee and Mukherjee 1994). Kastori et al. (1992) reported higher proline content under lead, cadmium, copper and zinc stressed crops. The proline accumulation helps to conserve nitrogenous compounds and protect the plant against heavy metal stress and acts as a membrane stabilizing agent under stress conditions (Poschenrieder and Barcelo 2004). Besides increase in proline content, enhancement of catalase, peroxidase and polyphenol oxidase activity are also the indicators of metal stress and the measurement of activity of these enzymes may reveal the invisible injuries caused to plants under metal stress. Sodium fluoride disturbs the seed germination, seedling growth and membrane stability and increases proline and carbohydrates in the seedlings. Similarly significant reduction in root length, shoot length, dry weight, fresh weight, germination, protein content, catalase activity, tolerance index, vigour index, germination rate, germination relative index, mean daily germination of gram seeds were observed at increasing fluoride concentration (Datta et al. 2012). The influence of NaF (0–30 μM) was studied on the germination, seedling growth and total biomass of the clusterbean under controlled conditions (Sabal et al. 2006) and reductions in seed germination, root and shoot length and total biomass were observed with increasing concentration of fluoride. The 30 μM NaF concentration was the lethal dose and 100 % mortality of the seeds was recorded. Increased chromium concentration in plants adversely affects several biological parameters of the plant leading to the loss of vegetation (Dube et al. 2003). The photosynthetic pigments (Kumar et al. 2004), nitrogen metabolism (Kumar and Joshi 2008), nutritive values and protein content are adversely affected under chromium (VI) stress (Kumar et al. 2010). Chromium treatment adversely affected the nitrogenase, nitrate reductase, nitrite reductase, glutamine synthetase and glutamate dehydrogenase activities in various plant organs at different growth stages and the specific activities of these enzymes decreased with increase in chromium (VI) levels. A concentration of 4 mg chromium (VI)/kg soil was found to be lethal to clusterbean plants (Sangwan et al. 2014a). The studies suggest that even 2 mg Cr (VI)/kg soil is toxic and has adverse effect on the nutritive value of forage clusterbean due to decreased protein content and increased structural carbohydrates (Sangwan et al. 2014b). Increased structural carbohydrates (NDF and ADF) leads to further decreased IVDMD and make it undesirable for use in animal feed.

Irrigation of the crop with polluted water has adverse effect on the seedling growth (Thakural and Kaur 1987), besides polluted water insecticides like monocrotophos gradually decreased germination, seedling growth, plant height, chlorophyll, biosynthesis and stomatal index in clusterbean (Ramulu and Rao 1987). The plant height was decreased due to chromium (Sharma and Sharma 1993), lead (Moustakas et al. 1994), nickel (Vijayarengan and Lakshmanachary 1995) and zinc (Kalyanaraman and Sivagurnathan 1993).

Although heavy metals at low concentrations may increase the dry matter yield of various plant parts (Krishna and Singh 1992), the higher concentrations of heavy metals, viz. cadmium, copper, zinc (Kalyanaraman and Sivagurunathan 1993) and nickel (Vijayarengan and Lakshmanachary 1994) reduced the dry matter yield of plants. Kaladharan and Vivekanandan (1990) assessed the photosynthetic potential and accumulation of assimilate in the developing chloroembryos of clusterbean and reported that with specific ions simple shading resulted in 65 % reduction in dry matter accumulation.

References

Abdel-Ghani AH (2009) Response of wheat varieties from semi-arid regions of Jordan to salt stress. J Agron Crop Sci 195:5–65

Adriano DC (1986) Trace elements in the terrestrial environment. Springer, New York

Aldasoro JJ, Matilla A, Nicolás G (1981) Effect of ABA, fusicoccin and thiourea on germination K+ and glucose uptake in chick-pea seeds at different temperatures. Physiol Plant 53:139–145

Anonymous (2014) Annual progress report. Central Arid Zone Research Institute, Jodhpur, p 45

Asher CJ, Bell RW, Edwards DG (1990) Growth and nodulation of tropical food legumes in dilute solution culture. Plant Soil 122:249–259

Ashraf MY, Akhtar K, Sarwar G et al (2002) Evaluation of arid and semi-arid ecotypes of guar (Cyamopsis tetragonoloba L.) for salinity (NaCl) tolerance. J Arid Environ 52:473–482

Bahr AA (2007) Effect of plant density and urea foliar application on yield and yield components of chickpea (Cicer arietinum). Res J Agric Biol Sci 3(4):220–223

Bala K, Rao AV, Tarafdar JC (1989) Occurrence of VAM associations in different plant species of the Indian desert. Arid Soil Res Rehabil 3:391–396

Bell RW, Edward BG, Asher CJ (1989) External calcium requirements for growth and nodulation of six tropical food legumes grown in flowering culture solution. Aust J Agric Res 40:85–96

Bhadoria RBS, Tomar RAS, Khan H et al (1997) Effect of phosphorus and sulphur on yield and quality of cluster bean (Cyamopsis tetragonoloba). Indian J Agron 42:131–134

Bhatia IS, Nagpal ML, Singh P et al (1979) Chemical nature of the pigment of the seedcoat of guar (Cluster Bean Cyamopsis tetragonolobus L. Taub). J Agric Food Chem 27:1274–1276

Bhattacharjee S, Mukherjee AK (1994) Influence of cadmium and lead on physiological and biochemical responses of Vigna unguiculata (L.) Walp. Seedlings. I. Germination behaviour, total protein, proline content and protease activity. Poll Res 13(3):269–277

Bibi A, Shakir A, Sadaqat HA et al (2014) Assessment of genetic association among seedling traits in guar (Cyamopsis tetragonoloba L.) genotypes under water stress conditions. Int J Res Stud Biosci 2(8):20–29

Boyer JS (1982) Plant productivity and environment. Science 218:443–448

Brooks A (1986) Effects of phosphorus nutrition on ribulose-1, 5-biphosphate carboxylase activation, photosynthetic quantum yield and amounts of some Calvin cycle metabolites in spinach leaves. Aust J Plant Physiol 13:221–237

Brown JC, Ambler JE, Chaney RL et al (1972) Differential responses of plant genotypes to micronutrients. In: Mortved JJ, Giordano PM, Lindsay WL (eds) Micronutrients in Agriculture, Soil Sci Soc Amer Inc. Madison, USA, pp 389–418

Burman U, Garg BK, Kathju S (2004) Interactive effects of thiourea and phosphorus on clusterbean under water stress. Biol Plant 48(1):61–65

Burman U, Garg BK, Kathju S (2009) Effect of phosphorus application on clusterbean under different intensities of water stress. J Plant Nutr 32(4):668–680

Cachorro P, Ortiz A, Cerda A (1994) Implications of calcium nutrition on the repose of *Phaseolus vulgaris* L. to salinity. Plant Soil 159(2):205–212

Cakmak I (1997) Role of potassium in protecting higher plants against photo-oxidative damage. In: Johnston AE (ed) Food security in the WANA region, the essential need for balanced fertilization. International Potash Institute, Basel, Switzerland, pp 345–352

Carlson JB (1973) Morphology. In: Caldwell BE (ed) Soybean improvement, production and uses. American Society of Agronomy, Wisconsin

Clijsters H, Van Assche F (1985) Inhibition of photosynthesis by heavy metals. Photosynth Res 7:31–40

Datta KS, Dayal J (1988) Effect of salinity on germination and early seedling growth of guar (*Cyamopsis tetragonoloba*). Indian J Plant Physiol 31:357–363

Datta JK, Maitra A, Mondal NK et al (2012) Studies on the impact of fluoride toxicity on germination and seedling growth of gram seed (*Cicer arietinum* L. cv. Anuradha). J Stress Physiol Biochem 8(1):194–202

Doijode SD (1989a) Deteriorative changes in cluster bean seeds in different conditions. Veg Sci 16:89–92

Doijode SD (1989b) Effect of temperatures and packaging on longevity of clusterbean seeds. Die Garten bauwissenschaft 54:176–178

Dube B, Tewari K, Chatterjee J et al (2003) Excess Cr alters uptake and translocation of certain nutrients in citrullus. Chemosphere 53:1147–1153

Ellis RH, Hong TD, Roberts EH (1987) The development of desiccation tolerance and maximum seed quality during seed maturation in six grain legumes. Ann Bot 59:23–29

Elsheikh EAE, Ibrahim KA (1999) The effect of Bradyrhizobium inoculation on yield and seed quality of guar (*Cyamopsis tetragonoloba* L.). Food Chem 65:183–187

Farooq M, Wahid A, Kobayashi N et al (2009) Plant drought stress: effects, mechanisms and management. Agron Sustain Dev 28:185–212

Flowers TJ (2004) Improving crop salt tolerance. J Exp Bot 55:307–319

Garg BK, Burman U (2001) Physiological aspects of drought and salt tolerance in clusterbean. Forage Res 27(1):1–10

Garg BK, Kathju S, Vyas SP et al (1986) Effect of saline waters on drought affected clusterbean. Proc Indian Acad Sci (Plant Sci) 96:531–538

Garg BK, Kathju S, Vyas SP et al (1996) Relative effects of sodium salts on growth, yield and metabolism of arid legumes. Plant Physiol Biochem 23:148–152

Garg BK, Kathju S, Vyas SP et al (1997a) Sensitivity of clusterbean to salt stress at various growth stages. Indian J Plant Physiol 2:49–53

Garg BK, Kathju S, Vyas SP et al (1997b) Alleviation of sodium chloride induced inhibition of growth and nitrogen metabolism of clusterbean by calcium. Biologia Plant 39:395–401

Garg BK, Kathju S, Vyas SP et al (1998) Influence of water deficit stress at various growth stages on some enzymes of nitrogen metabolism and yield in clusterbean genotypes. Indian J Plant Physiol 3:214–218

Garg BK, Kathju S, Vyas SP et al (2003) Relative performance evapotranspiration, water and nitrogen use efficiency and nitrogen metabolism of clusterbean genotypes under arid environment. In: Henry A, Kumar D, Singh NB (eds) Advances in arid legumes research. Scientific Publishers, Jodhpur, pp 212–220

Garg BK, Burman U, Kathju S (2005) Physiological aspects of drought tolerance in clusterbean and strategies for yield improvement under arid conditions. J Arid Legume 2(1):61–66

Garg BK, Burman U, Kathju S (2006) Influence of thiourea on photosynthesis, nitrogen metabolism and yield of clusterbean (Cyamopsis tetragonoloba L. Taub) under rainfed conditions of Indian arid zone. Plant Growth Regul 48(3):237–245

Goswami M, Naik ML (1992) Effect of fertilizer effluent on chlorophyll content of Cyamopsis tetragonoloba L. Taub. J Environ Biol 13:169–174

Grenway H, Muns R (1980) Mechanisms of salt tolerance in non halophytes. An Rev Plant Physiol 31:149–190

Gutierrez-Boem F, Thomas GW (1999) Phosphorus nutrition and water deficits in field grown soybeans. Plant Soil 207:87–96

Harrington JK (1973) Biochemical basis of seed longevity. Seed Sci Technol 1:453–461

Jat G, Bagdi DL, Kakralya BL (2012) Mitigation of salinity induced effects by using Brassinolide in clusterbean. Bhartiya Krishi Anushandhan Patrika 27(1):18–22

Jerlin R, Srimathi P, Vanangamudi K (2001) Seed physiological and maturity indices. In: Vanangamudi K, Bharathi A, Natesan P et al (eds) Recent techniques and participatory approaches on quality seed production. Department of Seed Science and Technology, Tamil Nadu Agricultural University, Coimbatore

Kadian N, Yadav K, Aggarwal A (2014) Interactive effect of arbuscular mycorrhizal fungi and potassium on growth and yield in Cyamopsis tetragonoloba (L.) under water stress. Researcher 6(8):86–91

Kaladharan P, Vivekanandan M (1990) Photosynthetic potential and accumulation of assimilate in the developing chloroembryos of Cyamopsis tetragonoloba L. Taub Plant Physiol 92:408–416

Kalavathi D, Ramamoorthy K (1992) A note on the effect of seed size on viability and vigour of seed in clusterbean (Cyamopsis tetragonoloba L. Taub.) cultivar Pusa Navbahar. Madras Agric J 79:530–532

Kalavathy D, Vanangamudi K (1990) Seed size, seedling vigor and storability in seeds of cluster beans. Madras Agric J 77:39–40

Kalyanaraman SB, Sivagurunathan P (1993) Effect of cadmium, copper and zinc on growth of blackgram. J Plant Nutr 16:2029–2042

Kant S, Kafkafi U (2002) Potassium and abiotic stresses in plants. In: Pasricha NS, Bansal SK (eds) Role of potassium in nutrient management for sustainable crop production in India. Potash Research Institute of India, Gurgaon, Haryana, pp 233–251

Kastori R, Petrovic M, Petrovic N (1992) Effect of excess lead, cadmium, copper and zinc on water relations in sunflower. J Plant Nutr 15(11):2427–2439

Kathju S, Aggarwal RK, Lahiri AN (1987) Evaluation of diverse effects of phosphate application on legumes of arid areas. Trop Agric 64:91–96

Khandelwal A, Sindhu SS (2013) ACC deaminase containing rhizobacteria enhance nodulation and plant growth in clusterbean (Cyamopsis tetragonoloba L.). J Microbiol Res 3(3):117–123

Khattra S, Singh G (1995) Indices of physiological maturity of seeds: a review. Seed Res 23(1):13–21

Khurana AL, Saini ML, Jhorar BS et al (1978) Nodulation studies in clusterbean. Forage Res 4:195–198

Krishna S, Singh RS (1992) Effect of zinc on yield, nutrient uptake and quality of Indian mustard. J Ind Soc Soil Sci 40:321–325

Kuhad MS, Sheoran IS (1986) Physiological and biochemical changes in clusterbean genotypes under water stress. Indian J Plant Physiol 29:46–52

Kumar D (2005) Breeding for drought resistance. In: Ashraf M, Harris PJC (eds) Abiotic stress: plant resistance through breeding and molecular approaches. Haworth Press, New York, pp 145–175

Kumar S, Joshi UN (2008) Nitrogen metabolism as affected by hexavalent chromium in sorghum (Sorghum bicolor L.). Environ Exp Bot 64:135–144

Kumar V, Kaushik MK (2014) Growth substances response to clusterbean (Cyamopsis tetragonoloba L.) growth parameters. Agric Sustain Dev 2(1):11–13

Kumar A, Sharma BK (1989) Note on salinity induced changes in galactomannan gum content in endosperm of guar seed. Adv Plant Sci 2:300–301

Kumar S, Joshi UN, Luthra YP (2004) Effect of Cr (VI) levels on plant growth and photosynthetic pigments in forage sorghum. Forage Res 29:176–179

Kumar S, Joshi UN, Sangwan S (2010) Cr (VI) influenced nutritive value of forage sorghum (*Sorghum bicolor* L.). Anim Feed Sci Technol 160:121–127

Kumar A, Kharb RPS, Mishra PK et al (2014) Studies on effect of priming treatments on germination and seedling establishment and their correlation in guar (*Cyamopsis tetragonoloba* L.). Forage Res 40(2):71–76

Kumawat RN, Dayanand, Mahla HR (2015) Effect of foliar applied urea and planting pattern on the leaf pigments and yield of clusterbean (*Cyamopsis tetragonoloba* L.) grown in low rainfall areas of western India. Legume Res 38(1):96–100

Lagerwerff JV, Specht AW (1970) Contamination of road side soil and vegetation with Cd, Ni, Pb and Zn. Sci Tech 7:583–586

Lahiri AN, Kachar NL (1985) Influence of potassium on plants under soil water deficit. Pr II Res Rev Ser 2:57–66

Lahiri AN, Garg BK, Kathju S et al (1987) Responses of clusterbean to soil salinity. Ann Arid Zone 26:33–42

Lepp NW (1981) Effects of heavy metal pollution on plants, vol 1. Applied Science Publishers, London

Lindhauer MG (1987) Solute concentrations in well watered and water stressed sunflower plants differing in K nutrition. J Plant Nutr 10:1965–1973

Liu W, Peffley EB, Powell RJ et al (2007) Association of seed coat color with seed water uptake, germination, and seed components in guar (*Cyamopsis tetragonoloba* (L.) Taub). J Arid Environ 70:29–38

Lohan SK, Jain S (2007) Improvement in colour purity and seed quality of clusterbean seed lot by digital colour sorter. Forage Res 33(2):91–94

Madalgeri MS, Rao MM (1989) Effect of fertilizer levels and staggered removal by early produced fresh pod on the quality of Pusa Navbahar clusterbean (*Cyamopsis tetragonoloba*) seeds. Seeds Farms 15:7–10

Manchanda HR, Sharma SK, Malik HR et al (1988) Technology for using sodic water for guar and bajra. Indian Farming 38:11–15

Manivasagaperumal R, Balamurugan S, Thiyagarajan G et al (2011) Effect of zinc on germination, seedling growth and biochemical content of cluster bean (*Cyamopsis tetragonoloba* (L.) Taub). Current Bot 2(5):11–15

Meena VK, Kaushik MK, Meena RS et al (2014) Effect of growth regulators on clusterbean [*Cyamopsis tetragonoloba* (L.)] growth under Aravali hills environment in Rajasthan. Bioscan 9(2):547–550

Mehta BV, Desai RS (1958) Effect of soil salinity on germination of some seeds. J Soil Water Conserv India 6:169–176

Mishra BK, Yadav V, Vishal MK et al (2013) Physiological and molecular characterization of clusterbean (*Cyamopsis tetragonoloba* (L.) Taub) rhizobia isolated from different areas of Rajasthan, India. Legume Res 36(4):299–305

Mittler R, Blumwald E (2010) Genetic engineering for modern agriculture: challenges and perspectives. Annu Rev Plant Biol 61:443–462

Mocquot B, Vangronsveld J, Clijestres J et al (1996) Copper toxicity in young maize (*Zea mays* L.) plants: effects on growth, mineral, chlorophyll contents and enzyme activities. Plant Soil 182(2):287–300

Moustakas M, Lanaras T, Symeonidis L et al (1994) Growth and some photosynthetic characteristics of field grown *Avena sativa* under copper and lead stress. Photosynthetica 30(3):389–396

Mpepereki S, Makonese F, Wollum AG (1997) Physiological characterization of indigenous rhizobial nodulation Vigna unguiculata in Zimbabwean soils. Symbiosis 22:275–292

Munns R, Tester M (2008) Mechanisms of salinity tolerance. Annu Rev Plant Biol 59:651–681

Murillo-Amador B, Troyo-Dieguez E, Lopez-Aguilar R et al (2002) Matching physiological traits and ion concentrations associated with salt tolerance with cowpea genotypes. Aust J Agric Res 53:1243–1255

Musil AF (1946) The germination of guar (*Cyamopsis tetragonoloba* (L.) Taub.). J Am Soc Agron 38:661–662

Nag P, Paul AK, Mukherjee S (1981) Heavy metal effects in plant tissues involving chlorophyll, chlorophyllase, hill reaction activity and gel electrophoresis patterns of soluble proteins. Indian J Exp Biol 19:702–706

Nandwal AS, Dabas S, Bharti S et al (1990) Zinc effect on nitrogen fixation and clusterbean yield. Ann Arid Zone 29:99–103

Neelkantan V, Mehta BV (1961) Studies on copper deficiency and toxicity symptoms in some common crops of Gujarat (India). J Agric Sci 56:293–298

Oke OL (1967) Nitrogen fixing capacity of guar beans. Trop Sci 9:144–147

Pankhrust CR, Sprent JI (1975) Effects of water stress on the respiratory and nitrogen fixing activity of soybean root nodules. J Exp Biol 26:287–304

Paredes-Lopez O, Calvino EC, Gonzalez-Castaneda J (1989) Effect of the hardening phenomenon on some physico-chemical properties of common bean. Food Chem 31:225–236

Pilbeam DJ, Cakmak I, Marschner H et al (1993) Effect of withdrawal of phosphorus on nitrate assimilation and PEP carboxylase activity in tomato. Plant Soil 154:111–117

Poschenrieder C, Barcelo J (2004) Water relations in heavy metal stressed plants. In: Prasad MNV (ed) Heavy metal stress in plants. Narosa Publishing House, New Delhi, India, pp 249–270

Prasad DDK, Prasad ARK (1987) Effect of lead and mercury on chlorophyll synthesis in mungbean seedlings. Phytochemistry 26:881–883

Rajan MR, Prema M, David NS (2014) Field level study on the impact of zinc electroplating industry effluent residue on growth, biochemical characteristics and yield of clusterbean (*Cyamopsis tetragonoloba*). Int J Sci Res 3(6):165–166

Rama Rao N (1986) Potassium nutrition of pearl millet subjected to moisture stress. J Potassium Res 2:1–12

Ramulu CA, Rao D (1987) Effect of monocrotophos on seed germination, growth and leaf chlorophyll content of cluster bean. Comparative Physiol Ecol 12:102–105

Ramulu CA, Rao D (1992) Influence of Dianthane M-45 on growth, flowering, fruiting and seed yield in guar (*Cyamopsis tetragonoloba* (L.) Taub.). Adv Plant Sci 5:12–16

Rao AV (1995) Biological nitrogen fixation in arid ecosystem. In: Behl RK, Khurana AL, Dogra RC (eds) Plant microbe interactions in sustainable agriculture. Bioscience Publishers, Hisar, pp 89–101

Rao AV, Venkateswarlu B (1983) Pattern of nodulation and nitrogen fixation in moth bean. Indian J Agric Sci 53:1035–1038

Rao AV, Venkateswarlu B, Henry A (1984) Genetic variation in nodulation and nitrogenase activity in guar and moth. Indian J Genet Plant Breed 44(3):425–428

Rathore SS, Pruthi NK, Rathore GPS (2009) Alleviation of sodium chloride stress by growth regulators in seedlings of clusterbean (*Cyamopsis tetragonoloba* L. Taub). Seed Res 37(1/2):40–47

Renugadevi J, Natarajan N, Srimathi P (2006) Studies on seed development and maturation in clusterbean (*Cyamopsis tetragonoloba*). Madras Agric J 93:195–200

Reynolds M, Tuberosa R (2008) Translational research impacting on crop productivity in drought-prone environments. Curr Opin Plant Biol 11:171–179

Sabal D, Khan TI, Saxena R (2006) Effect of sodium fluoride on cluster bean (*Cyamopsis tetragonoloba*) seed germination and seedling growth. Res Rep Fluoride 39(3):228–230

Sangwan P, Kumar V, Joshi UN (2014a) Effect of Chromium (VI) toxicity on enzymes of nitrogen metabolism in clusterbean (*Cyamopsis tetragonoloba* L.). Enzyme. doi:10.1155/2014/784036

Sangwan P, Kumar V, Joshi UN (2014b) Chromium (VI) affected nutritive value of forage clusterbean (*Cyamopsis Tetragonoloba* L.). Int J Agric Environ Biotechnol 7(1):17–23

Sathiyamoorthy L, Vivekanandan M (1995) Vigour of *Cyamopsis tetragonoloba* (guar) seed produced from foliar spray of nitrogenous salts under rainfed conditions. Legume Res 18:50–52

Schlegel H, Godbold DL, Hutchinson A (1987) Whole plant aspects of heavy metal induced changes in CO_2 uptake and water relations of spruce (*Picea abies*) seedlings. Physiol Plantarum 69:265–270

Serraj R, Adu-Gyamfi J (2004) Role of symbiotic nitrogen fixation in the improvement of legume productivity under stressed environments. West Afr J App Ecol 6:95–109

Seth R, Padmavathi C (2003) Expression of seed vigour and hard seededness in relation to seed coat colour in fodder clusterbean (*Cyamopsis tetragonoloba* (L.) Taub.) cultivars. Forage Res 28(4):185–189

Sharma SJ, Lashkari CO (2009) Response of gibberellic acid and cycocel on growth and yield of clusterbean (*Cyamopsis tetragonaloba* L.) cv. 'Pusa Navbahar'. Asian J Hortic 4(1):89–90

Sharma DC, Sharma CP (1993) Chromium uptake and its effects on growth and biological yield of wheat. Cereal Res Commun 21(4):317–322

Sharma AK, Parashar BN, Sharma CR (2014) Impact of paper mill effluent on seed germination and seedling growth of *Cyamompis Tetragonoloba* L. in Sanganer region of Jaipur (Rajasthan), India. J Environ Sci Comput Sci Eng Technol 3(4):1830–1835

Sheoran IS, Khan MI, Garg OP (1980) Differential behavior of guar genotypes to water stress during germination. Guar Newsletters 1:3–4

Sheoran IS, Luthra YP, Kuhad MS et al (1982) Clusterbean: a ureide or amide producing legume. Plant Physiol 70:917–928

Sheoran IS, Gupta VKI, Laura JS et al (1991) Whole plant aspects of heavy metal induced changes in CO_2 uptake and water relations in pigeon pea (*Cajanus cajan* L.) exposed to excess cadmium. Int J Exp Biol 29:857–861

Shubhra DJ (2005) Impact of phosphorus application on leaf characteristics, nodule growth and plant nitrogen content under water deficit in clusterbean. Forage Res 31(3):212–214

Shubhra, Dayal J, Goswami CL et al (2004) Influence of phosphorus application on water relations, biochemical parameters and gum content in clusterbean under water deficit. Biol Plant 48 (3):445–448

Singh M, Lal P, Singh KS (1976) A preliminary study on the inducement of salt tolerance in guar (*Cyamopsis tetragonoloba*) var. Chikna at germination stage. Agrochem 20:88–92

Sivakumar P, Rajan MR (2014) Effect of zinc electroplating industry effluent residue on growth and certain biochemical characteristics of clusterbean (*Cyamopsis tetragonoloba* L. Taub.). Int J Sci Res 3(9):368–370

Skirycz A, Inze D (2010) More from less: plant growth under limited water. Curr Opin Biotechnol 21(2):197–203

Stafford RE, Lewis CR (1980) Nodulation in inoculated and non-inoculated Kinman guar. A and M University of Agricultural Research and Extension Centre, Texas, USA

Stafford RE, McMichael BL (1990) Primary root and lateral root development in guar seedlings. Environ Exp Bot 30(1):27–34

Stobart AK, Griffiths WT, Ameen-Bukhari I et al (1985) The effect of Cd^{2+} on the biosynthesis of chlorophyll in leaves of barley. Physiol Plantarum 63:293–298

Terry N (1981) Physiology of trace element toxicity and its relation to iron stress. J Plant Nutr 3:561–578

Thakural AK, Kaur P (1987) Effect of some trace elements of polluted water on the germination of *Cyamopsis tetragonoloba* L. Taub. Indian J Ecol 14:185–188

Umar S, Moinuddin (2002) Genotypic differences in yield and quality of groundnut as affected by potassium nutrition under erratic rainfall conditions. J Plant Nutr 25:1549–1562

Uprety M, Yadav RBL (1985) Influence of growth retardant (B-9) on growth, flowering, fruiting and seed yield in guar plants under pot culture. Indian J Plant Physiol 28:1282–1289

Venkatesh MS, Basu PS (2011) Effect of foliar application of urea on growth, yield and quality of chickpea under rainfed conditions. J Food Legumes 24(2):110–112

Venkateswarlu B, Rao AV, Lahiri AN (1983) Effect of water stress on nodulation nitrogenase activity of guar (*Cyamopsis tetragonoloba* (L.) Taub.). Proc Indian Acad Sci (Plant Sci) 92(3):297–301

Verma CH, Sankhla N (1976) Effect of ethephon on growth and metabolism of *Cyamopsis tetragonoloba* (L.) Taub. Forage Res 2:41–48

Vijayarengan P, Lakshmanachary AS (1994) Differential nickel tolerance in greengram cultivars. Poll Res 13(3):291–296

Vijayarengan P, Lakshmanachary AS (1995) Effects of nickel on growth and dry matter yield of greengram cultivars. Indian J Environ Health 37(2):99–106

Vyas SP, Kathju S, Garg BK et al (1985) Performance and metabolic alterations in *Sesamum indicum* L. under different intensities of water stress. Ann Bot 56:323–333

Vyas SP, Garg BK, Kathju S et al (2001) Influence of potassium on water relation, photosynthesis, nitrogen metabolism and yield of clusterbean under soil moisture deficit stress. Indian J Plant Physiol 6:30–37

Weaver RW, Arayangkool T, Schomberg HH (1990) Nodulation and nitrogen fixation of guar at high root temperature. Plant Soil 126:209–213

Yadav NK, Vyas SR (1971) Response of root nodules rhizobia in saline alkaline and arid conditions. Indian J Agric Sci 41:875–881

Yadav BD, Joon RK, Singh A (2004) Effect of thiourea and kinetin on productivity of clusterbean under rainfed conditions. Forage Res 30(1):36–38

Zahran HH (1999) Rhizobium-legum symbiosis and nitrogen fixation under severe conditions and in an arid climate. Microbiol Mol Biol Rev 63:968–989

Zheng GH, Gu Zh, Xu Bh (1980) A physiological studies of germination in guar (*Cyamopsis tetragonoloba*). Acta Phytophysiol Sin 6:115–126

Chapter 7
Genetic Markers and Biotechnology

Abstract Clusterbean is an obligatory self-pollinating and inbreeding species and therefore, little variation in DNA polymorphism is expected among the cultivars. It has triploid endosperm and has more vegetative growth as compared to diploid counterparts. The endosperm tissue often shows a high degree of chromosomal variations, polyploidy, miotic irregularities, chromosome bridges and laggards. A brief detail of various techniques, viz. RAPD—random amplified polymorphic DNA; ISSR—inter simple sequence repeats; EST—expressed sequence tag markers; SCAR—sequence characterized amplified regions for amplification of specific band; CAPs—cleaved amplified polymorphic sequences; ITS—internal transcribed spacer; SSRs—Simple sequence repeats used for various purposes in clusterbean is summarized in this chapter.

7.1 Introduction

The information on the genetic diversity has several applications for crop improvement including reduction of number of accessions required to ensure broad range of genetic variability. The genetic diversity and relatedness between genotypes/ varieties within inter- and intra-species were initially assessed using qualitative and quantitative traits based on morphological characters. But these analyses are influenced by environmental factors and require number of years along with tedious statistical procedures (Pathak et al. 2011). So, the assessment of genetic diversity was shifted from morphological markers to the protein markers during early 1930s. These markers are useful in cultivar identification (Park et al. 2009), but are limited by the number of informative markers and do not give direct assessment of the potential variation existing in the genome and the profiling is influenced by tissue specificity and developmental stage. Molecular markers reveal natural variation at the DNA sequence level (Jones et al. 1997). These markers can be used to study the genetic diversity because of its potential to show variation at molecular level. Molecular markers are being widely used in various areas of plant breeding as an important tool for evaluating genetic diversity.

© Springer Science+Business Media Singapore 2015
R. Pathak, *Clusterbean: Physiology, Genetics and Cultivation*,
DOI 10.1007/978-981-287-907-3_7

Genetic marker reveals genetic variation in various traits, viz. phenotypic traits, allozymes, segments of the DNA and a gene within and among individuals and taxa. The markers and genes are close and stay together on the same chromosome of each generation of the plant and on the basis of these lineages, a genetic map can be prepared (Semagn et al. 2006). The genetic maps serve different purposes comprising associations between important traits and genes or quantitative traits loci and facilitate the introgression of desirable genes (Semagn et al. 2010). Various molecular markers can be classified into different groups based on nuclear inheritance, organelle inheritance, dominant or co-dominant markers and hybridization or polymerase chain reaction (PCR) based markers.

7.2 Biochemical Markers

Isozymes or isoenzymes are enzyme variants perform the same function but differ in amino acid sequence and refer to multiple forms of an enzyme sharing a catalytic activity derived from a tissue of a single organism. The indirect single-locus approach involves the utilization of protein which shows polymorphism and is derived from a single locus. Such proteins are generally known as isozymes. The term isozyme was introduced by Markert and Møller in (1959). Isozymes provided the first protein-based molecular marker system for genetic mapping in plants, and in the past, it has been the most widely used molecular markers (Weeden and Lamb 1985). It is still the quickest and cheapest marker systems to identify low levels of genetic variation.

Allozymes are co-dominant markers and two alleles at an allozyme locus in a heterozygous individual can be detected in multi-allelic manner (Baker 2000). With the discovery of allozyme techniques (Hunter and Markert 1957) and the allozyme variation within populations of Drosophila (Lewontin and Hubby 1966), the technique has been successfully applied to various related fields of genetic studies such as population genetics (Wright et al. 2003), plant systematics (van der Bank et al. 2001) and germplasm management (Chen et al. 2005). *Cyamopsis* isozyme diversity had been studied in relation to domestication of accessions (Mauria 2000). Brahmi et al. (2004) studied the allozyme diversity in clusterbean germplasm and found tris-borate system (pH 8.3) as the most effective system for detection of allozyme polymorphism in clusterbean. The clusterbean accessions from south-west Rajasthan showed higher variability supported by higher morphological variation in this region, while the genetic distance was low among the accessions, indicating its low diversity in India. Characterization of germplasm using biochemical techniques has received attention because of the increasing recognition of germplasm resources in crop improvement and in selection of desirable genotypes to be used in breeding programmes (Ghafoor and Ahmad 2005). The protocol includes starch gel, polyacrylamide gel electrophoresis but the protocol has to be standardized separately for isolation and purification of individual enzyme. The isozymic analysis is phenotype-based, and therefore, it may be influenced by environmental factors to distinguish closely related varieties/genotypes. Similarly, the biochemical methods are highly influenced by tissue specificity and developmental stages (Sammour 1991). Therefore, they may not qualify as co-dominant allozymes for use as genetic markers. Because of all these drawbacks, genetic marker analysis was shifted from use of protein markers to DNA-based marker in 1980.

7.3 DNA-Based Molecular Markers

DNA-based molecular markers analysis are important tools and have found their own position in various field like genetic engineering, taxonomy, physiology, embryology, etc. They are constantly being modified to enhance their utility in the process of genomic studies. The development of PCR is landmark in this effort and proved to be a unique process that brought about a new class of DNA profiling markers including development of marker based gene tags, map-based cloning of agronomically important genes, variability studies, phylogenetic analysis, synteny mapping, marker-assisted selection of desirable genotypes, etc. (Mullis and Faloona 1987). DNA markers allow direct assessment of variation in genotypes and their power of discrimination is so high that very closely related genotypes/varieties can also be distinguished. It is not affected by the growth, age, physiological condition of samples and environment effects and therefore, the basis of inheritance can be clearly understood. Only a small amount of sample is sufficient for analysis and its physical form is not any constraint. Jeffreys et al. (1985) introduced the term DNA fingerprinting first time to describe barcode-like DNA fragment patterns generated by multi-locus probes after electrophoresis separation of genomic DNA fragments.

In current scenario, the DNA-based marker has become the choice of markers for the study of crop genetic diversity in various areas of plant breeding for evaluating genetic diversity and variety/species identification (Agarwal et al. 2008) and the database help breeders to categorize the origin and degree of relatedness of land races, cultivars and varieties. These markers provide an opportunity to identify and isolate the gene regulating the character by map-based cloning and gives insight to create a favourable gene combination for producing a unique genotype that differ from their parents.

A number of molecular markers are utilized to evaluate DNA polymorphism and may be classified as hybridization-based markers, PCR-based markers and sequence-based markers (Agarwal et al. 2008). In hybridization-based markers, DNA profiles are visualized by hybridizing the restriction enzyme digested DNA to a DNA fragment of known origin or sequence. The hybridization-based marker assay is time-consuming having low polymorphism and requires radioactive isotopes along with large amount of DNA. They are comparatively slow, expensive and not responsive to automation. PCR is an extremely sensitive technique and operates at a very high speed. PCR-based markers comprise in vitro amplification of particular DNA sequences or loci specifically or arbitrarily chosen primers and a thermostable DNA polymerase enzyme. The sequence-based markers utilize the gene sequences and are the most recent marker systems.

7.4 PCR-Based Marker Techniques

PCR has got an enormous impact on molecular research due to its simplicity and robustness. It requires small amount of DNA, no prior knowledge and has ability to amplify DNA sequences even from preserved tissues. The technique is able to generate many genetic markers within a short time and to screen many genes simultaneously either as direct collection of data or as a feasibility study prior to nucleotide sequencing (Wolfe and Liston 1998). PCR-based different molecular markers techniques, viz. RAPD—random amplified polymorphic DNA; ISSR—inter simple sequence repeats; EST—expressed sequence tag markers; SCAR—sequence characterized amplified regions for amplification of specific band; CAPs—cleaved amplified polymorphic sequences; ITS—internal transcribed spacer; SSRs—Simple sequence repeats have been used for various purposes in clusterbean. A brief detail of these techniques are given below.

7.4.1 Random Amplified Polymorphic DNA (RAPD) Analysis

RAPD markers are amplification products of anonymous sequence obtained using single, short (8–10 bp) and oligonucleotide primers, and they do not require prior knowledge of DNA sequence. The technique was developed by Welsh and McClelland in 1991 and is a non-radioactive assay that requires only nanogram quantity of DNA. In this reaction, a single species of primer anneals to the genomic DNA at two different sites on complementary strands of DNA template and DNA product is formed through thermocyclic amplification under low annealing temperatures. RAPD markers have been recently used in studying the genetic relationships of clusterbean (Pathak et al. 2010, 2011; Rodge et al. 2012; Sharma et al. 2013, 2014a, b; Sharma and Sharma 2013; Kuravadi-Aswathnarayana et al. 2013; Ajit and Priyadarshani 2013; Kumar et al. 2013; Kalaskar et al. 2014; Patel et al. 2014).

Weixin et al. (2009) carried out genetic relationship among commercial cultivars of clusterbean using RAPD markers and found that *Cyamopsis tetragonoloba* is a distinct species from *Cyamopsis serrata* and *Cyamopsis senegalensis*. *C. senegalensis* was more related to the *C. serrata*, similar results were reported by Hiremath et al. (1996) who assessed the genetic relationships in the genus *Cyamopsis* using allozymes, RFLP and RAPD markers. Punia et al. (2009) characterized 18 genotypes of clusterbean into two groups based on RAPD markers and reported 0.34–0.76 % genotypic similarity coefficients among these genotypes. Pathak et al. (2010) carried out the molecular characterization of clusterbean genotypes using RAPD markers. The most distinct genotypes RGC-1002 and CAZG-6, exhibited the maximum genetic diversity of about 37 %. The maximum genetic similarity up to 98 % was observed between the genotypes RGC-2021 and CLBH-201. Using RAPD, Rodge et al. (2012) found no correlation between the diversity of the variety and its galactomannan content whereas, they reported that wild variety of clusterbean shows great diversity than the other varieties. Sharma and Sharma (2013) explored the correlation among RAPD and chemical markers in clusterbean and detected number of phenolic acids (sinapic acid, cholorogenic acid, caffeic acid and gallic acids) and flavonoids (kaempferol and myricetin). They suggested that the phytochemical analysis of clusterbean may expand its nutraceutical and pharmaceutical utilization. Kuravadi-Aswathnarayana et al. (2013) studied the genetic diversity of clusterbean landraces using RAPD and ISSR markers and found that the accessions belonging to central Rajasthan have grouped together showing the genetic similarity. Their results based on RAPD and RAPD + ISSR data grouped landraces and commercial varieties separately and showed the presence of distinguishable genetic difference between landraces and commercial varieties.

Pathak et al. (2010) correlated the molecular genetic diversity of clusterbean genotypes with their geographical locations and found that the genotypes from the same locations were grouped into different clusters revealing that the genotypes may have whole or partial common pedigrees and may have been subjected to the same selection during their breeding. A positive correlation between different genotypic varieties with respect to morphophysiological characteristics using RAPD analysis as well as in cultivar identification was designated in clusterbean lines (Punia et al. 2009). The analysis based upon agro-morphological, biochemical and molecular data revealed high genetic diversity among the clusterbean genotypes (Kumar et al. 2013).

Although RAPD is relatively fast, cheap and easy to perform but the issue of reproducibility has been of much concern with this technique since its inception. Several factors have been reported to influence the reproducibility of RAPD reactions including quality and quantity of template DNA, PCR buffer, concentration of magnesium chloride, primer-to-template ratio, annealing temperatures, Taq DNA polymerase, source and thermocycler (Wolff et al. 1993). These concerns about reproducibility, however could be overcome through choice of an appropriate DNA extraction protocol to remove any contaminants (Micheki et al. 1994), by optimizing the parameters (Skroch and Nienhuis 1995), by testing several oligonucleotide primers and scoring only the reproducible DNA fragments (Yang and Quiros 1993) and by using appropriate DNA polymerase, concentration of primer and template DNA (Muralidharan and Wakeland 1993) etc.

7.4.2 Inter Simple Sequence Repeats (ISSR) Analysis

It is a PCR-based multi-locus marker system that utilizes oligonucleotide primers for amplification of DNA segments present in between two similar microsatellite repeat regions oriented in opposite direction (Zietkiwicz et al. 1994). The technique uses microsatellite as primers in a single-primer PCR reaction targeting multiple genomic loci to amplify mainly inter simple sequence repeats of different sizes. The microsatellite repeats used as primers for ISSRs can be dinucleotide, trinucleotide, tetranucleotide or pentanucleotide. The primers used can either be unanchored or usually be anchored at 3′ or 5′ end with 1–4 degenerated bases extended into the flanking sequences (Meyer et al. 1993). The technique has much similarity with RAPD except that ISSR primer sequences are designed from microsatellite regions. The higher stringency of amplification in the form of longer primers (16–15 bp) and elevated annealing temperature depends on the GC content of the primer used and ranges from 45 to 65 °C. The amplified products are usually 200–2000 bp long and amenable for detection by both agarose and polyacrylamide gel electrophoresis (Joshi et al. 2000). The technique is simple, quick and does not require radioactive isotopes and shows high polymorphism (Kojima et al. 1998). It is advantageous over AFLPs in terms of high multiplex ratio and over RAPDs in terms of 3.5-fold greater variability (Zehdi et al. 2004) and as compared to RAPD, the risk of non-specific primer annealing is reduced.

Recently this marker technique has been used to detect DNA polymorphism and genetic diversity in clusterbean (Kuravadi-Aswathnarayana et al. 2013; Sharma et al. 2014a, b). Sharma et al. (2014a, b) used RAPD and ISSR markers to estimate genetic diversity and relationships among 35 clusterbean genotypes and found the equal importance of both the markers in diversity analysis. The GA-repeat and AC-repeat based primers produced more number of bands and higher polymorphism, whereas primers with AG-repeats produced comparatively less numbers of bands. Sharma et al. (2012) investigated the molecular and phytochemical methods for 35 genotypes of clusterbean and suggested that the markers generated by RAPD and ISSR can provide practical information for the management of genetic resources, molecular classification and breeding of new varieties.

7.4.3 Sequence Characterized Amplified Region (SCAR)

SCAR is a DNA fragment amplified by the PCR using specific 15–30 bp primers, designed from nucleotide sequences established in cloned RAPD/ISSR fragments linked to a trait of interest (Michelmore et al. 1991; Martin et al. 1991). This has led to a new generation of molecular markers for genetic determination. SCAR does not face problem of low reproducibility due to use of longer PCR primers and has better reproducibility than RAPDs/ISSRs. SCARs are usually dominant markers, however some of them can be converted into co-dominant markers by digesting them with tetra cutting restriction enzymes and polymorphism can be realized. The SCAR technique is simple, fast and relatively inexpensive and exhibit several advantages in mapping studies and map-based cloning. SCARs also allow comparative mapping or homology studies among related species.

The technique involves genomic DNA isolation, cloning and sequencing of RAPD/ISSR/AFLP marker bands and standardization of SCAR PCR conditions. Depending upon purpose, cytoplasmic or genomic DNA is isolated following standard DNA isolation protocols. The target DNA fragments in the RAPD/ISSR/AFLP reactions are extracted from agarose gels by standard gel purification techniques and the fragments are cloned into a plasmid vector. *Escherichia coli* cells are transformed with the recombinant vector and plated onto media plates. Positive colonies are determined and plasmids are randomly selected. The plasmids containing the fragments of the correct size are sequenced in an automated sequencer using forward and reverse primer. The DNA sequences are aligned from multiple clones using clustalW or clustalX software. The clones exhibiting single nucleotide polymorphisms (SNPs) or insertions/deletions on the basis of resistant versus susceptible varieties to major diseases and insect pests, physiological stresses, etc. are used to develop forward and reverse primers.

In the second step, PCR amplification is performed and a single band of interest appeared on agarose gel is isolated. About 10 μl of amplified DNA product is digested with 10 μl of a mixture containing 1 unit of endonucleases restriction enzyme (e.g. *Alu* I, *Rsa* I, *EcoR*-I, *Hinf* I, *Hind* III, *Mse* I, *Taq* I) in separate tubes along with specific buffers and the mixture is incubated at 37 °C for 1–2 h in case of most of the restriction enzymes except *Taq* I, which is incubated at 60 °C. Digested DNA of different samples is run on agarose gel or polyacrylamide gel electrophoresis in the presence of uncut control and λ DNA ladder, stained with ethidium bromide and photographed. This facilitates use of SCAR marker in quick identification of heterozygous alleles for a character of interest in unknown samples.

Sharma et al. (2014a, b) carried out the molecular characterization of clusterbean genotypes using RAPD, ISSR and SCAR markers system and observed wide range of diversity in both RAPD and ISSR markers. The polymorphic and geographical specific bands from RAPD as well as genotype-specific band from genotype RGC-1031 from ISSR were selected and converted into SCAR markers. The genotype-specific marker, i.e. SCAR-20 for RGC-1031 (tolerant genotype against *Macrophomina phaseolina*), could be used to prove identity of the genotype for improvement as well for its genetic purity assessment. Another SCAR-8 was selected due to its specificity for clusterbean genotypes from Rajasthan which might be important for population admixture studies.

7.4.4 Cleaved Amplified Polymorphic Sequence (CAPS)

The technique involves amplification of a segment of genomic DNA, cloning and sequencing of RAPD/ISSR/AFLP marker bands using specific primers and subsequent restriction digestion of the PCR product. The polymorphism can be studied by electrophoretic separation of the restriction fragments. The technique is also known as PCR-RFLP marker. The PCR primer for this purpose can be synthesized based on the sequence information available in databank of genomic or cDNA sequences or cloned RAPD bands. These are co-dominant markers in nature. Konieczny and Ausubel (1993) developed a set of CAPS markers and first adapted the CAPS procedure for genetic mapping in *Arabidopsis*. CAPS mapping procedure is technically simple, robust, requires small quantity of DNA and ensures amplification in every reaction. CAPS markers can be assayed relatively quickly using standard agarose gel electrophoresis. The technique has also been used to elucidate the genetic diversity in *M. phaseolina* that infects clusterbean (Purkayastha et al. 2006).

7.4.5 Expressed Sequence Tag (EST)

Expressed sequence tag is a unique stretch of DNA within a coding region of a gene. It utilizes reverse transcriptase enzyme to convert mRNA to complementary DNA (cDNA) and serves as a landmark for mapping. Since mRNA is highly unstable outside the cell, an enzyme called reverse transcriptase is used to convert mRNA to cDNA which is more stable and represents only expressed DNA sequence. After cDNA isolation, one can sequence a few hundred nucleotides either from 5′ or 3′ end to create 5′ expressed sequence tags and 3′ expressed sequence tags, respectively (Jongeneel 2000). The ESTs developed by transcriptome profiling can be used as a cost-effective source for the development of molecular markers such as SNPs and SSRs (Zhangying et al. 2011). ESTs provide an alternative to library screening, full-length cDNA sequencing and SNP data mining.

Naoumkina et al. (2007) described the features of EST dataset derived from single-pass sequencing of cDNAs of developing clusterbean seeds for understanding the seed-specific gene expression, by providing an extensive resource for the cloning of genes, development of markers for map-based cloning and annotation of genomic sequence information and reported the expression patterns of sets of genes involved in the storage of polysaccharide and protein metabolism during the development of clusterbean seeds. They suggested that the biosynthesis of carbohydrate and storage proteins in clusterbean seeds is probably due to the increased transcriptional activity. Naoumkina et al. (2007) constructed a database of 16,476 clusterbean seed ESTs from cDNA libraries and found that approximately 27 % of the clones were not similar to the known sequences due to the non-coding RNA, suggesting the uniqueness of the genes. Besides this they found high flux of energy into carbohydrate and storage protein synthesis in clusterbean seeds and identified involvement of unigenes in galactomannan metabolism. Dhugga et al. (2004) reported mannan synthase gene and identified it from an EST collection derived from RNA isolated from clusterbean seeds at three different stages of development.

7.4.6 Internal Transcribed Spacers (ITS)

ITS are region within the ribosomal transcript that are excised and degraded during maturation. The region used is usually a combination of ITS and ribosomal sequences and consists of 18S rRNA (or 16S rRNA for prokaryotes) sequence, ITS-1, the sequences of 5.8S rRNA, ITS-2 and a sequence of 28S rRNA. The 18S and 28S rRNA genes evolve relatively slowly and are useful in addressing broad phylogenetic hypothesis (Hillis and Dixon 1991). The direct DNA sequencing offers several advantages over cloning and direct RNA sequencing, (i) the method utilizes relatively crude preparations of total DNA, (ii) only small amounts of DNA are required for amplification, (iii) both strands of the gene can be sequenced which reduces errors and (iv) comparable to automated DNA sequencing that utilize fluorescently labelled sequencing primers or dNTPs. The ITS primers make use of conserved regions of the 18S, 5.8S and 28S rRNA genes to amplify the non-coding regions between them. Currently, nuclear ribosomal ITS is considered as one of the most useful phylogenetic marker due to less functional constrains and comparatively higher evolutionary changes. The length and sequences of ITS region of rDNA repeats are believed to be fast evolving and therefore may fluctuate (Cullings and Vogler 1998). The rate and patterns of ITS sequence mutation are typically appropriate for resolving relationships among species and genera. The nuclear rRNA gene complex is a tandem repeat unit of one to several thousand copies and has several domains that evolve at varying rates (Jorgenson and Cluster 1988) with different phylogenetic utilities. ITS has been the most reliable and useful technique for molecular systematics at the species level (Vogler and Bruns 1998) and sometimes even within the species (Pathak et al. 2010; Kakani et al. 2011).

The use of rDNA sequencing is limited for analysing genetic diversity in clusterbean. Pathak et al. (2011) amplified and sequenced the ITS regions of ribosomal DNA of five clusterbean genotypes using forward PCR primer ITS-1 (5'-TCC GTA GGT GAA CCT GCG G-3') and reverse primer ITS-4 (5'-TCC TCC GCT TAT TGA TAT GC-3') developed by White et al. (1990). They performed the PCR amplification in a total volume of 50 ml containing 1U Taq DNA polymerase, 2.5 mM $MgCl_2$, 160 mM dNTP mix, 50 pmol of each ITS-1 and ITS-4 primers, 50 ng genomic DNA in dH_2O. The reactions were performed in the gradient thermocycler for 34 cycles at 95 °C for 1 min, 50 °C for 30 s, and 72 °C for 1 min 20 s with a final elongation step of 72 °C for 10 min. The PCR products were subjected to electrophoresis in 1.6 % agarose gel. The gels with amplification fragments were visualized and photographed under UV light. The amplified ITS region containing ITS-1, 5.8S rDNA and ITS-2 were directly sequenced using forward primer of ITS-1 and reverse primer of ITS-4. The sequenced data of forward reverse primer were aligned using software to obtain the complete sequence of ITS region. The nucleotide sequence comparisons were performed by using BLAST network services available at the NCBI, USA database. The gene sequences (FJ769261-FJ769265) submitted, are the novel record for completed sequences of

this crop. The high frequencies of SNPs at seven sites in PCR-amplified products
of conserved gene region enabled to reveal close lineage of genetically distinct
genotypes, which can facilitate to transfer required character by conventional as
well biotechnological mean (Pathak et al. 2011).

7.4.7 Simple Sequence Repeats (SSRs)

Simple sequence repeats, also known as microsatellites, are tandem repeats of
2–6 bp DNA core sequences (Morgante et al. 2002) repeated in tandem and random manner. The repeated sequence is often simple, consisting of two, three or
four nucleotides such as $(AT)_{25}$, $(GAG)_{15}$, $(GACA)_{30}$ dispersed throughout the
genome and are highly polymorphic than other genetic markers. Such length-
polymorphisms can easily be detected using high resolution gel electrophoresis.
These markers often present high levels of inter- and intra-specific polymorphism,
particularly when the tandem repeat numbers are 10 or greater. The SSRs are efficient molecular markers due to their co-dominant inheritance, high allelic diversity
(Rakoczy-Trojanowska and Bolibok 2004) and can be identified by searching the
DNA data basis (Hares 2012).

Various approaches to identify and characterize SSRs have been developed and
used to identify functional SSR markers in different species. However, high development costs, labour-intensive process and development time are the major constraints for isolation of new SSR markers (Roder et al. 1998; Hayden and Sharp
2001). The locus-specific microsatellite-based markers have been reported from
many plant species (L'taief et al. 2012; Nie et al. 2012) and have wide applications in the field for cultivar identification, pedigree analysis, characterization of
germplasm diversity and genetic mapping studies. Microsatellite markers have
been applied widely in plant genetic studies for construction of linkage maps and
QTL mapping (Isemura et al. 2012) or evolution studies (Patto Vaz et al. 2004).
Kuravadi-Aswathnarayana et al. (2014) studied the mining of SSR sequences from
16,476 ESTs of clusterbean using the microsatellite identification tool and identified
907 SSR. The study has emerged with designing and validation of 187 EST-SSR
markers in five accessions of clusterbean, including three cultivated varieties of
C. tetragonoloba and two wild *Cyamopsis* species. The results suggest that approximately one SSR occurs per 4.1 kb of ESTs of clusterbean.

Each PCR is carried out in a about 25 ml reaction mixture containing approximately 15 ng of template DNA, nuclease free water, $10 \times$ buffer, 50 mM $MgCl_2$,
2–5 mM of dNTPs, 2–3 U of Taq polymerase, and 5 pmol of each forward and
reverse primers. The PCR amplification of genomic DNA is done by incubating
the DNA samples for 5 min at 94 °C, then 40–45 cycles comprising 94 °C for
60 s, annealing of primer for 60 s at 58–60 °C and the extension for 60 s at 72 °C.
The final extension is carried out for 10 min at 72 °C. The amplification products are separated in 6 % polyacrylamide gels under an initial voltage of 60 V for
30 min, extending it to 120 V for about 2 h in TBE buffer and observed under a

UV trans-illuminator. Standard molecular weight markers of 10 and 100 bp may be used. Bands are scored for absence and presence of bands. The cluster analysis is performed using the NTSYSpc software to determine genetic diversity and similarity (Rohlf 1992).

7.5 Callus Induction and Regeneration Protocols

The recalcitrant to transformation and in vitro regeneration of clusterbean is highly genotype-specific. One of the important constraints towards biotechnological manipulation of clusterbean is its poor ability to respond to the in vitro culture of endosperm, genetic transformation and subsequent regeneration of the transformed tissue. Only few reports are available on clusterbean plant tissue culture establishment and growth (Bansal et al. 1994; Prem et al. 2003, 2005; Ramulu and Rao 1993; Bhansali 2011). The endosperm tissue often shows a high degree of chromosomal variations, polyploidy, miotic irregularities, chromosome bridges and laggards. Clusterbean has triploid endosperm and have more vegetative growth as compared to diploid counterparts. An efficient regeneration system would facilitate transgenic regulation of the genes involved in galactomannan formation as well as other agronomically important genes in clusterbean.

Bansal et al. (1994) reported direct regeneration from cotyledon explants of clusterbean on Murashige and Skoog's (MS) medium supplemented with indole-3-acetic acid (IAA) or indole-3-butyric acid (IBA) and from hypocotyl explants on MS medium supplemented with IAA, 2,4-D or IBA. Ramulu and Rao (1991) reported embryoids formation from callus or leaf explants of clusterbean using 1-naphtaleneacetic acid and 6-benzylaminopurine, however, the development of these shoot/embryoids into plantlets and subsequent regeneration has not been explained. Prem et al. (2003) found high potential of cotyledonary nodes for direct shoot regeneration and plant regeneration from callus for several in vitro manipulations, viz. hybrid embryo rescue, in vitro mutagenesis and cell line screening. Prem et al. (2005) studied the effect of various growth regulators and their combinations on a variety of explants namely embryo, cotyledons, cotyledonary nodes, shoot tip and hypocotyle in clusterbean and developed an efficient system for callus induction and regeneration. The medium containing 2,4-D in combination with 6-benzylaminopurime (BAP) with embryo or cotyledon explants is found the most suitable for induction of callus in clusterbean. Bhansali (2011) found that the MS culture media containing 2,4-D, IAA, 1-napthaleneacetic acid (NAA) in combination with BAP with embryo or cotyledon seed explants is the most suitable for induction of dull white and friable endosperm callus. MS media containing 6-benzylaminopurine in combination with IAA with cotyledon node explants gives the highest frequency of multiple shoot regeneration (Prem et al. 2003). More efficient regeneration is reported following culturing callus on MS medium containing NAA in combination with BAP with a range of 82.1–88.4 % of callus clumps producing 20–25 shoots (Prem et al. 2005). Ramulu and Rao (1996) examined the response of four auxins 2,4-D, NAA, IAA and picloram in tissue culture of two genotypes of clusterbean and picloram was found the most effective auxins for callus growth of genotypes, NAA concentrations gave good growth, IAA supported effective callus growth whereas, 2,4-D enhanced the callus growth. Deepika et al. (2014) observed the response of various explants from 7-day-old seedlings of clusterbean and found cotyledonary node as best explants and MS medium supplemented with vitamins B_5, 2,4-D and BAP as the best callus growth supporter.

Mathiyazhagan et al. (2013) observed the best response for callus induction on the combination of media 2,4-D + BAP in case of cultivated clusterbean cultivars whereas, in case of wild species, higher concentration of 2,4-D + BAP had the best response. Ahlawat et al. (2013) studied the in vitro callus formation in cultivated and wild species of *Cyamopsis* and found that 2,4-D and BAP induced callusing from cotyledons in the species of *Cyamopsis*. They reported maximum callus induction from cotyledon explant in *C. serrata* and *C. senegalensis* on a medium supplemented with 2,4-D whereas, *C. tetragonoloba* showed poor callus formation on the same medium. The callus however, proliferated well on MS medium adjuncted with NAA + BAP. When the concentration of NAA was increased to 1 mg/litre and concentration of BAP was decreased with 0.5 mg/l, response of callus induction was decreased in *C. senegalensis* while no change response was found in *C. serrata* and *C. tetragonoloba*. Verma et al. (2013) formulated a novel combination of plant growth regulators comprising of IBA, BA and GA_3 in MS basal medium for in vitro induction of shoot and root in single culture using cotyledonary node explants of clusterbean. They reported highest percentage of shoot and root induction in the medium containing 2, 3 and 1 mg/l of IBA, BA and GA_3 respectively. Shoot and root regeneration was observed after 10–15 and 20–25 days, respectively along with successful acclimatization on transfer to soil. The genotypic differences observed in callus initiation response of various clusterbean genotypes at different media compositions clearly indicate that callus induction is genetically controlled traits.

For tissue culture, the seeds are washed thoroughly with tap water for 5–10 min and subsequently, surface sterilization is done with 70 % alcohol for 1 min followed by 0.1 % mercuric chloride solution for 5 min. The seeds are then washed carefully three to four times in sterile distilled water to remove all traces of mercury. The sterilized seeds are germinated on a medium containing 2–3 % sucrose and 7–8 % agar under aseptic conditions. The process should initially be done under dark conditions until germination and then it should be shifted to light conditions. The explant, viz. cotyledon, cotyledonary node and hypocotyl measuring 4 to 5 mm is obtained aseptically from the grown seedlings and is inoculated on the surface of different culture medium. The germination medium may contain MS (Murashige and Skoog 1962) salts, sucrose and bacteriological-grade agar supplemented with B_5 vitamins fortified with different concentrations of growth regulators. The callus induction medium may contain MS salts supplemented with different concentration of BAP, IAA, 2,4-D, IBA and/or NAA. In case of embryo explants, it is excised from 10-day-old green pods, surface sterilized and three explants per flask are cultured. Inoculated flasks were kept in culture room at 25 ± 1 °C temperature, under photoperiod of 16 h light and 8 h darkness. The developed plantlets are taken out of the culture vessels, washed to remove the sticking medium and planted in plastic pots for hardening. The pots should be watered with 0.5–0.6 % bavistin solution. On successful adaptation, the plantlets may be transplanted to soil. The callus culture of clusterbean endosperm can also be done.

Ahmad and Anis (2007) developed an efficient in vitro regeneration procedure using thidiazuron (TDZ) to permit multiple shoot induction from cotyledonary node explants of clusterbean. Shoot bud induction occurred on MS medium after 4 weeks in the presence of TDZ, followed by transfer onto shoot multiplication and elongation media containing MS salts, B_5 vitamins and different combinations of auxins and cytokinins. The combinations of auxins and cytokinins showed stimulatory effect on shoot multiplication and also on the length of the newly formed shoots. They reported maximum shoot induction on agar-solidified medium containing 5 μM BA with 0.5 μM IAA. Rooting of in vitro regenerated shoots was achieved in ex vitro conditions by a pulse treatment with 300 μM IBA. Rooted plantlets were transferred to soil where 70–75 % plants attained sexual maturity and produced viable seeds under greenhouse conditions. Gargi et al. (2012) reported cotyledonary node and cotyledon as the most appropriate explants amenable to callus induction with the highest pooled average callus induction frequency with 2–4-D and BAP for the cotyledons based on a total of 200 explants whereas, the hypocotyls were the least responsive to callus induction in clusterbean.

The desire to increase the viscosity of the gum of clusterbean has led to the development of chemical derivatives. The galactomannan is synthesized in the endosperm of seed by the co-action of two membrane-bound enzymes, viz. mannan synthase and galactosyl transferase. The mechanism of their synthesis has been studied in *Trigonella foenum-graecum* and clusterbean (Reid and Bewley 1979; Reid et al. 2003; Edwards et al. 2004). The activity for these two enzymes attains the peak level in the endosperm at around 25–35 days after flowering (Dhugga et al. 2004; Naoumkina et al. 2007). Clusterbean endosperm is the main target for improving the physical properties of galactomannan. Development of suitable endosperm-specific promoters for use in clusterbean is always desirable for metabolic engineering of the seed gum because the promoter plays the most important role in determining the temporal and spatial expression pattern of a gene. Presently some strong constitutive promoters, viz. cauliflower mosaic virus 35S promoter, maize ubiquitin promoter, etc. are widely used (Saidi et al. 2009) for studying expression patterns. But the expression level of the target gene in the desired tissue is often not satisfactory (Drakakaki et al. 2000). The continuous high expression of a foreign gene in all tissues or in non-target tissues may cause harmful effects in the host plant (Cheon et al. 2004). The bean phaseolin promoter is expressed in both embryo and endosperm, use of this promoter to drive mannan synthase (MS) transgene increased the length of galactomannan chain in clusterbean (Naoumkina et al. 2008). Using a strong endosperm-specific promoter to restrict only the endosperm gene expression is the important solution in this direction. The 35S promoter is the most commonly used promoter for driving transgene expression in plants. Isolation of an endogenous endosperm-specific promoter is enviable for genetic engineering of galactomannan metabolism in clusterbean to prevent off-target genetic and metabolic distresses in the plant. The promoter of the MS gene is suitable for this purpose as this gene is highly expressed in the endosperm tissue of clusterbean seeds. Dhugga et al. (2004) isolated the first gene (CtManS) from clusterbean seeds that encodes a mannan synthase and demonstrated how to encode a functional mannan synthase enzyme in transformed soybean somatic embryos. These soybean embryos, which normally have no detectable mannan synthase activity, exhibited significant mannan synthase activity when the CtManS cDNA sequence was overexpressed.

Naoumkina and Dixon (2011) isolated 1.6 kb MS promoter region in clusterbean and fused its sequence with the GUS reporter gene and overexpressed in alfalfa (*Medicago sativa*) and found that the MS promoter directs GUS expression specifically endosperm in transgenic alfalfa. Thus, the MS promoter of clusterbean could be useful for directing endosperm-specific expression of transgenes in other legume species. Joersbo et al. (1999) studied the transmission of transgene in clusterbean and found that the major parts of the transgenes displayed aberrant transmission at a high frequency in the process of the primary transformants to the first offspring generation. The method for clusterbean transformation was optimized to function irrespective of the cultivar but the transformation frequencies varied somewhat. Genetically modified plants must transmit their transgenes faithfully through subsequent generations (Fearing et al. 1997). Silencing of transgenes, epidermal transformants, transgene instability, loss or deletions of transgenes are the major reasons for the abnormal transgene expression.

Molecular documentation of clusterbean germplasm is an urgent need for conservation and assessment of variability intended for morphophysiological and industrial qualities because genetically diverse lines provide substantial opportunity to create a gene combination and probability for producing a unique genotype having higher yield and gum content.

References

Agarwal M, Shrivastava N, Padh H (2008) Advances in molecular marker techniques and their applications in plant sciences. Plant Cell Rep 27:617–631

Ahlawat A, Dhingra HR, Pahuja SK (2013) In vitro callus formation in cultivated and wild species of *Cyamopsis*. Afr J Biotechnol 12(30):4813–4818

Ahmad N, Anis M (2007) Rapid plant regeneration protocol for cluster bean (*Cyamopsis tetragonoloba* L. Taub.). J Hortic Sci Biotechnol 82(4):585–589

Ajit P, Priyadarshani Y (2013) Molecular characterization of cluster bean (*Cyamopsis tetragonoloba*) cultivars using PCR-based molecular markers. Int J Adv Biotechnol Res 4(1):1021–1029

Baker AJ (ed) (2000) Molecular methods in ecology. Blackwell Science Ltd., USA

Bansal YK, Chibbar T, Bansal S et al (1994) Plant regeneration from hypocotyls and cotyledon explants of clusterbean (*Cyamopsis tetragonoloba*). J Physiol Res 7:57–60

Bhansali RR (2011) Callus culture from endosperm of clusterbean (*Cyamopsis tetragonoloba* L. Taub). J Arid Legumes 8(2):77–82

Brahmi P, Bhat KV, Bhatnagar AK (2004) Study of allozyme diversity in guar (*Cyamopsis tetragonoloba* L. Taub.) germplasms. Genetic Res Crop Evol 51:735–746

Chen J, Wang PS, Xia YM et al (2005) Genetic diversity and differentiation of *Camellia sinensis* L. (cultivated tea) and its wild relatives in Yunnan province of China, revealed by morphology, biochemistry and allozyme studies. Genet Resources Crop Evol 52:41–52

Cheon BY, Kim HJ, Oh KH et al (2004) Overexpression of human erythropoietin (EPO) affects plant morphologies: retarded vegetative growth in tobacco and male sterility in tobacco and Arabidopsis. Transgenic Res 13:541–549

Cullings KW, Vogler DR (1998) A 5.8S nuclear ribosomal RNA gene sequence database: application to ecology and evolution. Mol Ecol 7:919–923

Deepika, Dhingra HR, Goyal SC (2014) Callus induction and regeneration from various explants in guar (*Cyamopsis tetragonoloba* (L.) Taub.). Indian J Plant Physiol 19(4):388–392

Dhugga KS, Barreiro R, Whitten B et al (2004) Guar seed β-mannan synthase is a member of the cellulose synthase super gene family. Sci 303:363–366

Drakakaki G, Christou P, Stoger E (2000) Constitutive expression of soybean ferritin cDNA in transgenic wheat and rice results in increased iron levels in vegetative tissues but not in seeds. Transgenic Res 9:445–452

Edwards ME, Choo TS, Dickson CA et al (2004) The seeds of *Lotus japonicus* lines transformed with sense, antisense, and sense/antisense galactomannan galactosyl transferase constructs have structurally altered galactomannans in their endosperm cell walls. Plant Physiol 134:1153–1162

Fearing PL, Brown D, Vlachos D et al (1997) Quantitative analysis of CryIA(b) expression in *Bt* maize plants, tissues, and silage and stability of expression over successive generations. Mol Breed 3:169–176

Gargi T, Acharya S, Patel JB et al (2012) Callus induction and multiple shoot regeneration from cotyledonary nodes in clusterbean (*Cyamopsis tetragonoloba* L. Taub.). AGRES-An Int e-J 1(1):1–7

Ghafoor A, Ahmad Z (2005) Diversity of agronomic traits and total seed protein in black gram *Vigna mungo* (L.) Hepper. Acta Biol Cracoviensia Ser Bot 47(2):69–75

Hares DR (2012) Expanding the CODIS core loci in the United States. Forensic Sci Int Genet 6:e52–e54

Hayden MJ, Sharp PJ (2001) Targeted development of informative microsatellite (SSR) markers. Nucleic Acids Res 29:e44. doi:10.1093/nar/29.8.e44

Hillis DM, Dixon MT (1991) Ribosomal DNA: molecular evolution and phylogenetic inference. Q Rev Biol 66:411–453

Hiremath SC, Ramamoorthy J, Cai Q et al (1996) Analysis of genetic relationships in the genus *Cyamopsis* (Fabaceae) using allozymes, RFLP and RAPD markers. Am J Bot 83:207

Hunter R, Markert C (1957) Histochemical demonstration of enzymes separated by zone electrophoresis in starch gels. Sci 125:1294–1295

Isemura T, Kaga A, Tabata S et al (2012) Construction of a genetic linkage map and genetic analysis of domestication related traits in mungbean (*Vigna radiata*). PLoS ONE 7(8):e41304. doi:10.1371/journal.pone.0041304

Jeffreys AJ, Wilson V, Thein SL (1985) Hypervariable 'minisatellite' regions in human DNA. Nature 314:67–73

Joersbo M, Brunstedt J, Marcussen J et al (1999) Transformation of the endospermous legume guar (*Cyamopsis tetragonoloba* L.) and analysis of transgene transmission. Mol Breed 5:521–529

Jones N, Ougham H, Thomas H (1997) Markers and mapping: we are all geneticists now. New Phytol 137:165–177

Jongeneel CV (2000) Searching the expressed sequence tag (EST) databases: panning for genes. Briefings Bioinf 1:76–92

Jorgenson RD, Cluster PD (1988) Modes and tempos in the evolution of nuclear ribosomal DNA: new characters of evolutionary studies and new markers for genetic and population studies. Ann Mo Bot Gard 75:1238–1247

Joshi SP, Gupta VS, Aggarwal RK et al (2000) Genetic diversity and phylogenetic relationship as revealed by intersimple sequence repeat (ISSR) polymorphism in genus *Oryza*. Theor Appl Genet 100:1311–1320

Kakani RK, Singh SK, Pancholy A et al (2011) Assessment of genetic diversity in *Trigonella foenum-graecum* based on nuclear ribosomal DNA, internal transcribed spacer and RAPD analysis. Plant Mol Biol Rep 29:315–323

Kalaskar SR, Acharya S, Patel JB et al (2014) Genetic diversity assessment in clusterbean (*Cyamopsis tetragonoloba* (L.) Taub.) by RAPD markers. J Food Legumes 27(2):92–94

Kojima T, Nagaoka T, Noda N et al (1998) Genetic linkage map of ISSR and RAPD markers in Einkorn wheat in relation to that of RFLP markers. Theor Appl Genet 96:37–45

Konieczny A, Ausubel FM (1993) A procedure for mapping *Arabidopsis* mutations using recombinant ecotype specific PCR based markers. Plant 4:403–410

Kresivich S, Williams JGK, McFerson JR et al (1992) Characterization of genetic identities and relationships of *Brassica oleracea* L. via a random amplified polymorphic DNA (RAPD) assay. Theor Appl Genet 85:190–196

Kumar S, Joshi UN, Singh V et al (2013) Characterization of released and elite genotypes of guar (*Cyamopsis tetragonoloba* (L.) Taub.) from India proves unrelated to geographical origin. Genet Res Crop Evol 60:1573–S109

Kuravadi-Aswathnarayana N, Tiwari PB, Choudhary M et al (2013) Genetic diversity study of cluster bean (*Cyamopsis tetragonoloba* (L.) Taub) landraces using RAPD and ISSR markers. Int J Adv Biotechnol Res 4(4):460–471

Kuravadi-Aswathnarayana N, Tiwari PB, Tanwar UK et al (2014) Identification and characterization of EST-SSR markers in clusterbean (*Cyamopsis* spp.). Crop Sci 54:1097–1102

L'taief B, Horres R, Jungmann R et al (2012) Locus-specific microsatellite markers in common bean (*Phaseolus vulgaris* L.): isolation and characterization. Euphytica 162(2):301–310

Lewontin RC, Hubby JL (1966) A molecular approach to the study of genetic heterozygosity in natural populations. II. Amount of variation and degree of heterozygosity in natural populations of *Drosophila pseudoobscura*. Genet 54:595–609

Markert CL, Møller F (1959) Multiple forms of enzymes: tissue ontogenetic and species specific patterns. Proc Natl Acad Sci (Wash) 45:753–756

Martin GB, Williams JGK, Tanksley SD (1991) Rapid identification of marker linked to a Pseudomonas resistance gene in tomato by using random primers and near isogenic lines. Proc Natl Acad Sci USA 88:2336–2340

Mathiyazhagan S, Pahuja SK, Ahlawat A (2013) Regeneration in cultivated (*Cyamopsis tetragonoloba* L.) and wild species (*C. serrata*) of guar. Legume Res 36(2):180–187

Mauria S (2000) Isozyme diversity in relation to domestication of gaur (Cyamopsis tetragonoloba (L.) Taub.). Indian J Plant Genet Res 13:1–10

Meyer W, Michell TG, Freedman EZ et al (1993) Hybridization probes for conventional DNA fingerprinting used as single primers in polymerase chain reaction to distinguish strains of *Cryptococcus neoformans*. J Clin Microbiol 31:2274–2280

Micheki MR, Bova R, Pascale E et al (1994) Reproducible DNA fingerprinting with the random amplified polymorphic DNA (RAPD) method. Nucleic Acids Res 22:1921–1922

Michelmore RW, Paran I, Kesseli RV (1991) Identification of markers linked to disease resistance gene bulked segregant analysis: a rapid method to detect markers in specific genomic region using segregation population. Proc Natl Acad Sci USA 88:9828–9832

Morgante M, Hanafey M, Powell W (2002) Microsatellites are preferentially associated with non-repetitive DNA in plant genomes. Nat Genet 30:194–200

Mullis KB, Faloona FA (1987) Specific synthesis of DNA in vitro via a polymerase catalysed reaction. Methods Enzymol 255:335–350

Muralidharan K, Wakeland EK (1993) Concentration of primer and template qualitatively affects products in random amplified polymorphic DNA PCR. Bio Tech 14:362–364

Murashige T, Skoog F (1962) A revised medium for rapid growth and bioassay with tobacco tissue cultures. Physiol Plant 10:949–966

Naoumkina M, Dixon RA (2011) Characterization of the mannan synthase promoter from guar (*C. tetragonoloba*). Plant Cell Rep 30:997–1006

Naoumkina M, Torres-Jerez T, Allen S et al (2007) Analysis of cDNA libraries from developing seeds of guar (*C. tetragonoloba* (L.) Taub.). BMC Plant Biol 7:62

Naoumkina M, Vaghchhipawala S, Tang Y et al (2008) Metabolic and genetic perturbations accompany the modification of galactomannan in seeds of *Medicago truncatula* expressing mannan synthase from guar (*Cyamopsis tetragonoloba* L.). Plant Biotechnol J 6:619–631

Nie X, Li B, Wang L et al (2012) Development of chromosome-arm-specific microsatellite markers in *Triticum aestivum* (Poaceae) using NGS technology. Am J Bot 99(9):e369–e371

Park Y, Lee JK, Kim N (2009) Simple sequence repeat polymorphisms (SSRPs) for evaluation of molecular diversity and germplasm classification of minor crops. Mol 14:4546–4569

Patel KA, Patel BT, Shinde AS et al (2014) Optimization of DNA isolation and PCR protocols for RAPD analysis in clusterbean (*Cyamopsis tetragonoloba* (L.) Taub). Can. J Plant Breed 2(2):43–46

Pathak R, Singh SK, Singh M et al (2010) Molecular assessment of genetic diversity in clusterbean (*Cyamopsis tetragonoloba*) genotypes. J Genet 89(2):243–246

Pathak R, Singh SK, Singh M (2011) Assessment of genetic diversity in clusterbean based on nuclear rDNA and RAPD markers. J Food Legumes 24(3):180–183

Patto Vaz MC, Pêgo S, Satovic Z et al (2004) Assessing the genetic diversity of Portuguese maize germplasm using microsatellite markers. Euphytica 137:63–72

Prem D, Singh S, Gupta PP et al (2003) High-frequency multiple shoot regeneration from cotyledonary nodes of guar (*Cyamopsis tetragonoloba* L. Taub.). In vitro Cell Dev Biol 39:384–387

Prem D, Singh S, Gupta PP et al (2005) Callus induction and de novo regeneration from callus in guar (*Cyamopsis tetragonoloba*). Plant Cell, Tissue Organ Cult 80:209–214

Punia A, Yadav R, Arora P et al (2009) Molecular and morphophysiological characterization of superior clusterbean (*Cyamopsis tetragonoloba*) varieties. J Crop Sci Biotech 12(3):143–148

Purkayastha S, Kaur B, Dilbaghi N et al (2006) Characterization of *Macrophomina phaseolina* the charcoal rot pathogen of clusterbean using conventional techniques and PCR based molecular markers. Plant Pathol 55:106–116

Rakoczy-Trojanowska M, Bolibok H (2004) Characteristics and a comparison of three classes of microsatellite-based markers and their application in plants. Cell Mol Biol Lett 9:221–238

Ramulu CA, Rao D (1991) Tissue culture studies of differentiation in a grain legume (*Cyamopsis tetragonoloba* (L.) Taub.). J Physiol Res 4(2):183–185

Ramulu CA, Rao D (1993) In vitro effect of phytohormones on tissue culture of clusterbean (*Cyamopsis tetragonoloba* L. Taub). J Physiol Res 20:7–9

Ramulu CA, Rao D (1996) Genotypic responses of clusterbean to auxins in tissue cultures. J Environ Biol 17(3):257–260

Reid JS, Bewley JD (1979) A dual role for the endosperm and its galactomannan reserves in the germinative physiology of fenugreek (*Trigonella foenum-graecum* L.), an endospermic leguminous seed. Planta 147:145–150

Reid JS, Edwards ME, Dickson CA et al (2003) Tobacco transgenic lines that express fenugreek galactomannan galactosyl transferase constitutively have structurally altered galactomannans in their seed endosperm cell walls. Plant Physiol 131:1487–1495

Roder MS, Korzun V, Wendehake K et al (1998) A microsatellite map of wheat. Genet 149:2007–2023

Rodge A, Jadkar R, Machewad G et al (2012) Studies on isolation, rheological properties and diversity analysis of guar gum. Res Plant Biol 2(5):23–31

Rohlf FJ (1992) The analysis of shape variation using ordinations of fitted functions. In: Sorensen JT, Foottit R (eds) Ordinations in the study of morphology, evolution and systematics of insects: applications and quantitative genetic rationals. Elsevier, Amsterdam, pp 95–112

Saidi Y, Schaefer DG, Goloubinoff P et al (2009) The CaMV 35S promoter has a weak expression activity in dark grown tissues of moss *Physcomitrella patens*. Plant Signal Behav 4:457–459

Sammour RH (1991) Using electrophoretic techniques in varietal identification, biosystematic analysis, phylogenetic relations and genetic resources management. J Islamic Acad Sci 4(3):221–226

Semagn K, Bjornstad A, Ndjiondjop MN (2006) An overview of molecular marker methods for plants. Afr J Biotech 5:2540–2568

Semagn K, Bjornstad Å, Xu Y (2010) The genetic dissection of quantitative traits in crops. Electron J Biotechnol 13(5). http://dx.doi.org/10.2225/vol13-issue5-fulltext-14

Sharma A, Sharma P (2013) Genetic and phytochemical analysis of Cluster bean (*Cyamopsis tetragonaloba* (L.) Taub) by RAPD and HPLC. Res J Recent Sci 2(2):1–9

Sharma A, Sharma P, Singh YP (2012) Phytochemical and genetic analysis of *Cyamopsis tetragonaloba* by RAPD, ISSR and HPLC. Biochem Cell Arch 12(1):31–40

Sharma A, Mishra S, Garg G (2013) Molecular characterization and genetic relationships among cluster bean genotypes based on RAPD analysis. Res J Pharma Biol Chem Sci 4(1):8–17

Sharma P, Kumar V, Venkat Raman K et al (2014a) A set of SCAR markers in clusterbean (*Cyamopsis tetragonoloba* L. Taub) genotypes. Adv Biosci Biotechnol 5:131–141

Sharma P, Sharma V, Kumar V (2014b) Genetic diversity analysis of clusterbean (*Cyamopsis tetragonoloba* (L.) Taub) genotypes using RAPD and ISSR markers. J Agric Sci Tech 16:433–443

Skroch P, Nienhuis J (1995) Impact of scoring error and reproducibility of RAPD data on RAPD based estimates of genetic distance. Theor Appl Genet 91:1086–1091

van der Bank H, van der Bank M, van Wyk B (2001) A review of the use of allozyme electrophoresis in plant systematics. Biochem Syst Ecol 29(5):469–483

Verma S, Gill KS, Pruthi V et al (2013) A novel combination of plant growth regulators for in vitro regeneration of complete plantlets of guar (*Cyamopsis tetragonoloba* (L.) Taub.). Indian J Exp Biol 51:1120–1124

Vogler DR, Bruns TD (1998) Phylogenetic relationships among the Pine stem rust fungi (*Cronartium* and *Peridermium* spp.). Mycologia 90:244–257

Weeden NF, Lamb RC (1985) Identification of apple cultivars by isozyme phenotypes. J Am Soc Hort Sci 110:509–515

Weixin L, Anfu H, Peffley EB et al (2009) Genetic relationship of guar commercial cultivars. Chin Agric Sci Bull 25(2):133–138

White TJ, Bruns SL, Taylor J (1990) Amplification and direct sequencing of fungal ribosomal RNA genes for phylogenetics. In: Innis MA, Geltand DH, Sninsky TT et al (eds) PCR protocols: a guide to methods and applications. Academic Press, San Diego, p 315

Wolfe AD, Liston A (1998) Contribution of PCR based methods to plant systematics and evolutionary biology. In: Soltis DE, Soltis PS, Doyle JJ (eds) Molecular systematics of plant II: DNA sequencing. Kluwer Academic Publishers, Berlin, pp 43–86

Wolff K, Schoen ED, Peter-Van RJ (1993) Optimizing the generation of random amplified polymorphic DNA in *Chrysanthemum*. Theor Appl Genet 86:1033–1037

Wright SI, Lauga B, Charlesworth D (2003) Subdivision and haplotype structure in natural populations of *Arabidopsis lyrata*. Mol Ecol 12:1247–1263

Yang X, Quiros C (1993) Identification and classification of celery cultivars with RAPD markers. Theor Appl Genet 86:205–212

Zehdi S, Sakka H, Rhouma A et al (2004) Analysis of Tunisian date palm germplasm using SSR markers. Afr J Biotechnol 3:215–219

Zhangying W, Jun L, Zhongxia L et al (2011) Characterization and development of EST-derived SSR markers in cultivated sweet potato (*Ipomoea batatas*). Plant Biol 11:1–9

Zietkiwicz E, Rafalski A, Labuda D (1994) Genomic fingerprinting by SSR anchored PCR amplification. Genomics 20:176–183

Annexure I

List of Split Manufacturers

Rajasthan

Adarsh Guar Gum, Udyog Barmer, Rajasthan
Annapurna Gum Industries, Fulasar, Sardarsahar, Churu, Rajasthan
Apex Gum and Chemical Industries, Bap, Phalodi Road, Jodhpur, Rajasthan
Arihant Gum Industries, Dhorimanna, Barmer, Rajasthan
Ashok Gum Industries, Industrial area, Bikaner Road, Nagaur, Rajasthan
Ashok Industries, Mini Industrial Area, Basani, Jodhpur, Rajasthan
Asso Gum Pvt. Ltd., Jaipur, Rajasthan
Banshi Gum Industries, Mini Industrial Area, Basani, Jodhpur, Rajasthan
Bhagvati Enterprises, Mini Industrial Area, Basani, Jodhpur, Rajasthan
Bhairav Guar Gum Industries, Gudamalani, Sanchor, Barmer, Rajasthan
Bhinmal Gum Pvt. Ltd., RIICO Industrial Area, Bhinmal, Rajasthan
Bohra Guar Gum Industries, Adarsh Industrial area, Barmer, Rajasthan
CP Industries, Khichan, Phalodi, Rajasthan
Chandak Industries, Nokha, Rajasthan
Chopra Guar Gum Industries, Mini Industrial Area, Basani, Jodhpur, Rajasthan
Desna Gum Industries, Nokha, Rajasthan
Devaashish Gum Pvt. Ltd., Bhinmal, Rajasthan
Dhoot Gum and Chemicals, Degana, Nagaur, Rajasthan
Gandhi Enterprises, E-92, Mini Industrial Area, II Phase, Basani, Jodhpur, Rajasthan
Gotam Gum and Chemicals, Sindhari, Barmer, Rajasthan
Goyal Gum Industries, Rani Bazar Industrial area, Bikaner, Rajasthan
HMB Industries, Nokha, Rajasthan, Hanumangarh Gum Industries, Nokha, Bikaner, Rajasthan
Hindustan Gum and Chemicals Ltd., E-283, Mini Industrial Area, Basani, Jodhpur, Rajasthan
International Guar Gum Industries, PWD, Rest House, Nohar, Rajasthan

© Springer Science+Business Media Singapore 2015
R. Pathak, *Clusterbean: Physiology, Genetics and Cultivation*,
DOI 10.1007/978-981-287-907-3

Jamboo Kumar Jain, Dall Mill, Mini Industrial Area, Basani, Jodhpur, Rajasthan
Jay Shree Gum Industries, Beechhwal Industrial area, Bikaner, Rajasthan
Jodhpur Dying and Bleaching Mill, Mini Industrial Area, Basani, Jodhpur, Rajasthan
Kankaria Gum and Chemials, Nokha, Bikaner, Rajasthan
Kishan Gum Industries, Nokha, Bikaner, Rajasthan
Laxmi Industries, E-17, Industrial area, Sumerpur, Rajasthan
Laxmi Udyog, Nokha, Bikaner, Rajasthan
Lodha Industries Ltd., Jodhpur, Rajasthan,
Lunawat Dall Mill, Mini Industrial Area, Basani, Jodhpur, Rajasthan
Lunawat Industries, Mathania, Rajasthan
Mahalaxmi Gum Industries, Sindari, Barmer, Rajasthan
Mahavir Gum and Chemicals, Sindri (Barmer), Rajasthan
Mahavir Gum Udyog, RIICO Industrial area, Barmer, Rajasthan
Maheshwari Gum Industries, Osian, Jodhpur, Rajasthan
Manidhari Guar Gum Industries, Siwda, Jalore, Rajasthan
Mardia Gum and Chemicals, Sanchor, Jalore, Rajasthan
Mehta Guar Gum Industry, Barmer, Rajasthan
Modi Gum Industries, Gantha ghar, Nohar, Churu, Rajasthan
Mohta Oil and Gum Industries, Industrial area, Jaisalmar, Rajasthan
Moondra Industries, Mini Industrial Area, Basani, Jodhpur, Rajasthan
Narayan Udyog, Near Railway Station, Sardar Sahar, Rajasthan
Navdeep Gum Industries, Sardar Sahar, Rajasthan
New Krishna Gum Industries, Sardar Sahar, Rajasthan
Nilkanth Polymers, Sardar Sahar, Rajasthan
Nokha Gum Industries, Nokha, Bikaner, Rajasthan
Nola Ram Dulichand Dal Mill, Sardar Sahar, Rajasthan
Pankaj Dal Mill, Industrial area, Phalodi, Rajasthan
Paras Gum Industries, Bikaner, Rajasthan
Prakash Gum Industies, Nokha, Rajasthan
RK Guar Gum Udyog, Dhorimanna, Barmer, Rajasthan
Radha Madhur Gum Industries, Churu, Rajasthan
Raghuvanshi Gum Industries, Bap, Phalodi, Jodhpur, Rajasthan
Rajasthan Gum Industries, F-1/2, Industrial area, Sikar, Rajasthan
Ravindra Udyog, Nokha, Rajasthan
SN Hisaria Gum Industries, Kanja Industrial area, Hanumangarh, Rajasthan
Sanjay Guar Gum Udyog, Chohtan, Barmer, Rajasthan
Satyam Exports, Mini Industrial Area, Basani, Jodhpur, Rajasthan
Shiv Jyothi gum industries, Industrial area, Hanumangarh, Rajasthan
Shiv Shankar gum industries, Sanchor, Jalore, Rajasthan
Shri Maheshwari Gum Industries, Mini Industrial Area, Basani, Jodhpur, Rajasthan
Shri Maheswari Gum Industries, Mini Industrial Area, Basani, Jodhpur, Rajasthan
Sumit Guar Gum Mills, Mini Industrial Area, Basani, Jodhpur, Rajasthan
Suraj Dal Mill, RIICO Industrial area, Barmer, Rajasthan

Thirupathi Trading Company, Near Railway Station, Sardar Sahar, Rajasthan
Vikas Gum and Chemicals, Shree Ganga Nagar, Rajasthan
Vikram Gum Industries, Sanchor—Jalore, Rajasthan
Vimal Guar Gum Mills, Mini Industrial Area, Basani, Jodhpur, Rajasthan
West Raj gum udyog, Opposite Krishi Mandi, Industrial area, Barmer, Rajasthan

Gujarat

Amba Gum Industries, 88/2, GIDC Estate, Phase-1, Vatva, Ahmedabad—382 445, Gujarat
Bipin Gum and Oil Mills, Market Yard, Bavla, Ahmedabad, Gujarat
Hindustan Gum and Chemicals Ltd., Birla Colony, Bhivani, Haryana
Indian Guar Gum Pvt. Ltd., Industrial Area, Hisar, Haryana
Jagat Industries Ltd., Gujarat Industries Development Corporation, Mehsana, Gujarat
Prakash Gum Industries Gujarat Industries Development Corporation, Deesa, Gujarat
Ramdev Agro Pvt. Ltd., Gujarat Industries Development Corporation, Mehsana, Gujarat
Shree Arbuda Gum Industries, Gujarat Industries Development Corporation, Dhanera Gujarat
Shree Krishna Gum Industries, Gujarat Industries Development Corporation, Deesa Gujarat
Shree Shanti Nath Gum Industries, Gujarat Industries Development Corporation, Deesa Gujarat
Subhalaxmi Gum Industries, Gujarat Industries Development Corporation, Deesa Gujarat

Haryana

Hindustan Gum and Chemicals Ltd., Birla Colony, Bhivani, Haryana
Indian Guar Gum Pvt. Ltd., Industrial Area, Hisar, Haryana
Jai Bharat Gum and Chemicals Ltd. Siwani, Haryana

Name and Addresses of Powder Manufacturers

Rajasthan

Arizona Gum and Chemical, T-7, Industrial Estate, Jodhpur, Rajasthan
Badar Enterprises, Phase II, Mini Industrial Area, Basani, Jodhpur, Rajasthan
Dabur (India) Ltd., Alwar, Rajasthan
Dhoot Dal Mills, Phase II, Mini Industrial Area, Basni, Jodhpur, Rajasthan
Digvijay Gumshem Pvt. Ltd., Osian, Jodhpur, Rajasthan
Dinesh Enterprises, E-274, Mini Industrial Area, Basani, Jodhpur, Rajasthan
Indian Gum Industries Ltd., B-5/7, Mini Industrial Area, Basani, Jodhpur, Rajasthan

Jaisons (India) Industries, F-547, II Phase, Mini Industrial Area, Basani, Jodhpur,
    Rajasthan
Jaisons (India) Industries, F-547, II Phase Mini Industrial Area, Basani, Jodhpur,
    Rajasthan
Lotus Gums and Chemicals, G 656/657, Mini Industrial Area, Basani, Jodhpur,
    Rajasthan
Lucid Colloids Limited, Jodhpur, Rajasthan
Naveen Gum and Chemicals, Plot No. G-142, Mini Industrial Area, Basani,
    Jodhpur, Rajasthan
Neelkanth Polymers, RIICO Industrial Area, Sardarshahr, Churu, Rajasthan
Rajasthan Chemical Industries, II Phase, Mini Industrial Area, Basani, Jodhpur,
    Rajasthan
Sarda Gums and Chemicals, Plant E-2, Industrial Area, Pali Marwar, Rajasthan
Satyam Enterprises, G-654-655, Mini Industrial Area, Basani, Jodhpur, Rajasthan
Shakti Industries, Dahizar Industrial area, Mandore, Jodhpur, Rajasthan
Shree Ram Gum and Chemicals, C-79, Marudhar Industrial area, Basni, Jodhpur,
    Rajasthan
Shrinath Gum and Chemicals, II Phase, Mini Industrial Area, Basani, Jodhpur,
    Rajasthan
Stanley Chemicals Pvt. Ltd., II Phase, Mini Industrial Area, Basani, Jodhpur,
    Rajasthan
Sunita Minechem Industries, E-394, Mini Industrial Area, Basani, Jodhpur,
    Rajasthan
Supreme Gums Pvt. Ltd., G-999/1000, Sitapura Industrial area (Extn), Jaipur,
    Rajasthan
Vikas WSP Ltd., B-86/87, Udyog Vihar, RIICO Industrial Area, Sri Ganganagar,
    Rajasthan

Gujarat

Abdulbhai Abdul Kader, 713, Phase IV, GIDC, Vatva, Ahmedabad
Adarsh Derivatives Ltd., 229, GIDC Indl. Estate, Chandisar, Gujarat
Altrafine Gums, 88/2, GIDC Estate, Phase 1, Vatva, Ahmedabad, Gujarat
Bhavani Chemicals, C1-91/15, GIDC Estate, Phase I, Vatva, Ahmedabad, Gujarat
H.B. Gum Industries Ltd., 11/B, Gujarat Industries Development Corporation,
    Kalol, Gujarat
Hindustan Gum and Chemicals Ltd., Viramgam,Near Ahmedabad, Gujarat
Indian Gum Industries Ltd., 227/233, Gujarat Industries Development Corporation,
    Naroda, Ahmedabad, Gujarat
Kalika Gums, Naroda, Ahmedabad, Gujarat
Premson Industries, Gujarat Industries Development Corporation, Vatva, Ahmedabad,
    Gujarat
Shashvat Gum Industries, Gujarat Industries Development Corporation, Vatva,
    Ahmedabad, Gujarat

Shrijee Chemical Industries, C1-91/7, Gujarat Industries Development Corporation, Vatva, Ahmedabad, Gujarat

Shubhlaxmi Industries, 49, Gujarat Industries Development Corporation, Deesa Banaskantha, Gujarat

Haryana

Hindustan Gum and Chemicals Ltd., Birla Colony, Bhivani, Haryana

Modern Gum and Chemicals, 52, Near Anaj Mandi, Bhiwani, Haryana

Source: Field Survey and internet sites.

Annexure II

List of Plant and Machinery Suppliers for Processing Of Guar Gum

Mumbai

Ami Associates and Consultants P. Ltd. (Project Consultants), 13, Manoj Industrial Estate, 40-A, Katrak Road, Wadala, Mumbai

Hifab Engineers Pvt. Ltd., D-148, Bonanza Industrial Estate, A, Chekravarti Road, Kandivali (E), Mumbai

Jasubhai Richard Simson, D-222/2 TTC Industrial Area, Thane Belapur Road, Nerul, New Mumbai

Satguru Industries, 11, Phahuja Industrial Estate, Saki Naka, Mumbai

New Delhi

Promivac Engineers, 37575, PNB Building, NS Marg, Darya Gunj, New Delhi

Haryana

Septu India Pvt. Ltd., 545/15, Basai Road, Bara Bazar, PB No. 4, Gurgaon, Haryana

Gujarat

Trumatic Engineers, 91/18, Gujarat Industries Development Corporation, Vatva, Ahmedabad

Rajasthan

Vijaylaxmi Enterprises, B-388-F, Mini Industrial Area, Basani, Jodhpur, Rajasthan

Reed Medway, Sector 4 Plot No. SE, Ballbgarh

Source: Field Survey.

© Springer Science+Business Media Singapore 2015
R. Pathak, *Clusterbean: Physiology, Genetics and Cultivation*,
DOI 10.1007/978-981-287-907-3

Printed in the United States
By Bookmasters